CAMBRIDGE LIBRARY COLLECTION

Books of enduring scholarly value

Mathematics

From its pre-historic roots in simple counting to the algorithms powering modern desktop computers, from the genius of Archimedes to the genius of Einstein, advances in mathematical understanding and numerical techniques have been directly responsible for creating the modern world as we know it. This series will provide a library of the most influential publications and writers on mathematics in its broadest sense. As such, it will show not only the deep roots from which modern science and technology have grown, but also the astonishing breadth of application of mathematical techniques in the humanities and social sciences, and in everyday life.

Arithmetical Books
from the Invention of Printing to the Present Time

In the preface to this work, mathematician Augustus De Morgan (1806–71) claims that 'The most worthless book of a bygone day is a record worthy of preservation.' His purpose in writing this catalogue, published in 1847, was to provide an accurate record of the early history of publishing on arithmetic, but describing only those books which he had examined himself. He surveyed the library of the Royal Society, works in the British Museum, the wares of specialist booksellers, and the private collections of himself and his friends to compile a chronological list of books from 1491 to 1846 (the final book being a work of his own), giving bibliographical details, a description of the contents, and sometimes comments on the mathematics on display. De Morgan's *Formal Logic* and a *Memoir of Augustus De Morgan* by his widow are also reissued in the Cambridge Library Collection.

Cambridge University Press has long been a pioneer in the reissuing of out-of-print titles from its own backlist, producing digital reprints of books that are still sought after by scholars and students but could not be reprinted economically using traditional technology. The Cambridge Library Collection extends this activity to a wider range of books which are still of importance to researchers and professionals, either for the source material they contain, or as landmarks in the history of their academic discipline.

Drawing from the world-renowned collections in the Cambridge University Library and other partner libraries, and guided by the advice of experts in each subject area, Cambridge University Press is using state-of-the-art scanning machines in its own Printing House to capture the content of each book selected for inclusion. The files are processed to give a consistently clear, crisp image, and the books finished to the high quality standard for which the Press is recognised around the world. The latest print-on-demand technology ensures that the books will remain available indefinitely, and that orders for single or multiple copies can quickly be supplied.

The Cambridge Library Collection brings back to life books of enduring scholarly value (including out-of-copyright works originally issued by other publishers) across a wide range of disciplines in the humanities and social sciences and in science and technology.

Arithmetical Books
from the Invention
of Printing
to the Present Time

Being Brief Notices of a Large Number of Works
Drawn Up from Actual Inspection

AUGUSTUS DE MORGAN

CAMBRIDGE
UNIVERSITY PRESS

CAMBRIDGE
UNIVERSITY PRESS

University Printing House, Cambridge, CB2 8BS, United Kingdom

Cambridge University Press is part of the University of Cambridge.
It furthers the University's mission by disseminating knowledge in the pursuit of
education, learning and research at the highest international levels of excellence.

www.cambridge.org
Information on this title: www.cambridge.org/9781108070959

© in this compilation Cambridge University Press 2014

This edition first published 1847
This digitally printed version 2014

ISBN 978-1-108-07095-9 Paperback

This book reproduces the text of the original edition. The content and language reflect
the beliefs, practices and terminology of their time, and have not been updated.

Cambridge University Press wishes to make clear that the book, unless originally published
by Cambridge, is not being republished by, in association or collaboration with,
or with the endorsement or approval of, the original publisher or its successors in title.

ARITHMETICAL BOOKS.

LONDON:
PRINTED BY ROBSON, LEVEY, AND FRANKLYN,
Great New Street, Fetter Lane.

ARITHMETICAL BOOKS

FROM

THE INVENTION OF PRINTING TO THE PRESENT TIME

BEING

BRIEF NOTICES OF A LARGE NUMBER OF WORKS

DRAWN UP FROM ACTUAL INSPECTION

BY

AUGUSTUS DE MORGAN

OF TRINITY COLLEGE, CAMBRIDGE

SECRETARY OF THE ROYAL ASTRONOMICAL SOCIETY: FELLOW OF THE CAMBRIDGE PHILOSOPHICAL SOCIETY
AND PROFESSOR OF MATHEMATICS IN UNIVERSITY COLLEGE, LONDON.

———◆———

" Much surprised, no doubt, would the worthy man have been, had any one told him that two
hundred years after his death, when no man alive would think his ideas on the nature of mathematics worth a look, the absence of better materials would make his list of" arithmeticians " not
only valuable, but absolutely the only authority on several points."—Dublin Review, No. XLI.

LONDON

TAYLOR AND WALTON

BOOKSELLERS AND PUBLISHERS TO UNIVERSITY COLLEGE

28 UPPER GOWER STREET

———

1847

VERY REVEREND GEORGE PEACOCK, D.D.

DEAN OF ELY, LOWNDEAN PROFESSOR,

&c. &c. &c.

———◆———

MY DEAR SIR,

It never entered into my head till now to adorn the front of any book of mine with an eminent name : and the reason I take to be, that I have hitherto never chanced to write a separate work* upon any subject with which the name of one individual was especially associated in the minds of those who study it. But you are the only Englishman now living who is known, by the proof of publication, to have investigated both the scientific and bibliographical history of Arithmetic : and this compliment, be the same worth more or less, is your due, and would have been, though my knowledge of you had been confined to your writings. And it is the more cordially paid from the remembrance of nearly a quarter of a century of personal acquaintance, and of many acts of friendship on your part.

I have rather grown than made this catalogue. It never occurred to me to publish on the subject, till I found, on a casual review of what I had collected, that I could furnish from my own books a more extensive list than Murhard, Scheibel, Heilbronner, or any mathematical bibliographer of my acquaintance, has described *from his own inspection*. Knowing, from sufficient experience, the general inaccuracy and incompleteness of scientific lists, I therefore determined to do what I could towards the correction of both, by describing as many works as I could manage to see. From

---

* Had the regulations of the work in which it appeared permitted, it would have been most peculiarly appropriate to have inscribed my treatise on the *Calculus of Functions* to my friend Mr. Babbage.

the Royal Society's library, the stock of Mr. Maynard the mathematical bookseller, and my own collections, with a few from the
British Museum and the libraries of private friends, including three
or four of great rarity from yourself, I have accordingly compiled
the present catalogue. I have also given in the Index, in addition
to the names of the authors whom I have examined, those of all
whom I could find recorded as having written on the subject of
Arithmetic, whether as teachers, historians, or compilers of special
tables for aid in the main operations, independently of logarithms
and trigonometry.

A great number of persons are employed in teaching arithmetic
in the United Kingdom. In publishing this work, I have the hope
of placing before many of them more materials for the prevention
of inaccurate knowledge of the literature of their science than they
have hitherto been able to command, without both expense and
research. Your History, unfortunately for them, is locked up in
the valuable, but bulky and costly, Encyclopædia for which it was
written. I may have the gratification of knowing that some, at
least, of the class to which I belong, have been led by my catalogue
to make that comparison of the minds of different ages which is one
of the most valuable of disciplines, and without which the man of
science is not the man of *knowledge*.

The most worthless book of a bygone day is a record worthy of
preservation. Like a telescopic star, its obscurity may render it
unavailable for most purposes; but it serves, in hands which know
how to use it, to determine the places of more important bodies.

The effect of this work which would please me best, would be,
that the professed bibliographer should find it too arithmetical, and
that the student of the history of the science should find it too
bibliographical. I might certainly have entered more into the
methods of the several works, with advantage to the reader. But I
could not attempt to write the complete annals of arithmetic; this
would require still more books: neither could I, without losing
sight of my plan altogether, combine the information here presented with that derived from other sources. That plan has been,
to attempt some rectification of the numerous inaccuracies of existing catalogues, by recording only what I have seen myself. And
it is a sufficient justification of the course I have taken, that I have
produced in this way a *larger* catalogue than yet exists among
those devoted, in whole or in part, to this particular subject.

None but those who have confronted the existing lists with the works they profess to describe, know how inaccurate the former are; and none but those who have tried to make a catalogue know how difficult it is to attain common correctness. There is now a prospect of this country possessing in time *such a record of books as can be safely consulted in aid of the history of literature*,—I refer to the intended catalogue of the library in the British Museum. I, for one, can only hope that the chance will not be lost by any attempt to expedite its formation, in deference to the opinion of those who either are not aware how bad existing lists are, or are willing to take more than a chance of having nothing better. If, through negligence or fear on the part of those who have really compared book-lists and books, the expression of public feeling which any *primâ facie* case against public officers so easily obtains, should succeed in hurrying the execution of this national undertaking, the result will be one more of those magazines from which non-existing books take their origin, and existing ones are consigned to oblivion by incorrect description. However extensive the demand for spoiled paper may be, it should be remembered that the supply is immense, and that there is no need to insist on Parliament, at least, furnishing a larger contingent than it is obliged to do already.

I make no apology for troubling you to read this diatribe : it is your affair, and mine, and that of every one who believes accuracy to be an essential characteristic of useful knowledge.

I remain,

My dear Sir,

Yours sincerely,

A. DE MORGAN.

UNIVERSITY COLLEGE, LONDON,
April 29, 1847.

_{*} In consequence of a little change in the plan of this work, made after the printing was commenced, there may be one or two places in which the reader must consult the *Introduction*, where he is told to consult the *Index*; and the *Additions*, where he is told to consult the *Introduction*.

INTRODUCTION.

AT the end of this work will be found an Index containing, besides the names of the authors mentioned in the Catalogue, who may be known by the paginal figures opposite, every name which I have found *any where* as belonging to a writer on arithmetic before 1800. Since that period I have not swelled the list from mere *catalogues*, but have contented myself with what I had either seen, or learnt from some other source. For the existence of such writers, or of their works, of course I do not vouch, except in those cases in which the work is found in the body of the Catalogue. Of those who stand unpaged in the index, all I can say is, that they are taken from many sources, and that there is in each case plenty of reason to make inquiry for their writings, and strong presumption that their works are still to be found.

A word with an asterisk before it is the name of a writer who is referred to by Dr. Peacock. The great apparent preponderance of German names arises partly from the number of works written on the subject in Germany being very considerable, partly from the bibliographical catalogues of that country being more full than those of any other, and partly from the names which seem to be German, really including Danes, Dutch, Belgians, and some Swiss. I am far from thinking that this list contains even the third part of the names of those who have really written on the subject. I have not been particular in searching for any thing after 1750, though I have not refused what came in my way. I might have very much increased the list of recent Germans, from Rogg's *Bibliotheca Selecta* (Tubingen, 1830). Looking at the various countries which enjoyed the art of printing from 1500 downwards, I have an impression, from all that I have gathered, which would lead me to suppose that the number of works on arithmetic published in Latin, French, German, Dutch, Italian,

b

Spanish, and English, up to the middle of last century, cannot be less than three thousand, which gives to each language
less than an average of one a year. Few of these would seem
to fall within the province of the historian. Dr. Peacock refers
to about a hundred and fifty. Unfortunately, history must of
necessity be written mostly upon those works which, by being
in advance of their age, have therefore become well known.
It ought to be otherwise, but it cannot be, without better preservation and classification of the minor works which people
actually use, and from which the great mass of those who
study take their habits and opinions. The *Principia* of Newton is, if we believe the title-page, a work of the seventeenth
century : but the account of the effect which it produced on
science belongs to the eighteenth. It was not till many years
after the publication of the *Principia*, that its predecessor in
doctrine, the great work of Copernicus, produced its full effect
upon general thought and habit. Nor have we any reason to
suppose that it could have been otherwise. The great exceptions will always bear, perhaps, as large a ratio to the average
power as ever they did : it is as likely as not, that if the intel-
ligence of the sixteenth century had been sufficient to verify
and receive the opinion of Copernicus at once, some predecessor of his might have been Copernicus, and he or another
of his day might have been Newton.

It is then essential to true history, that the minor and secondary phenomena of the progress of mind should be more
carefully examined than they have been. We must distinguish
between the progress of possibilities and that of actual occur-
rences. Our written annals shew us too much what might have
been, and too little what was : they give some words to the slow
reception of an improvement, and more sentences to the account of the one man who was able to make it before the world
at large could appreciate it.

The public is beginning to demand that civil history shall
contain something more than an account of how great generals
fought, great orators spoke, and great kings rewarded both
for serving their turn. The progress of nations might well be
described, for the most part, with much less mention of any
of the three: but the parallel does not hold of knowledge.
Copernicus and Newton would fill a large space, though the
history were written down to that of every individual who
ever opened a book : but it seems to me that they, and their
peers, are made to fill *all the space*. Nor will it be otherwise
until the historian has at his command a readier access to

second and third rate works in large numbers ; so that he may write upon effects as well as causes.

This list contains upwards of fifteen hundred names, of which a few may be duplicates of others, arising from the wrong spelling of my authorities.* But these must be much more than counterbalanced by the number of names which belong to two or more authors, but which only appear once in my list. Thus there are two Digges's, three Riese's, two named Le Gendre, Walker, Newton, Wallis, &c., and several Taylors, Butlers, &c., though each name is only mentioned once. There are several cases in which I have not ventured to strike out one of two names, though there is every reason to suppose one is a mistake for the other ; as Caraldus and Cataldus, Cappaus and Cuppaus, Döhren and Dühren. But I have often been deceived in this way ; and have more than once been obliged to re-insert names which I had struck out, supposing them to be only different spellings of names already in the list. There are several whom I have seen in more places than one, who are clearly Germans with names metamorphosed by wrong read- ing of the black letter, or use of the genitive for a nominative, or both. Thus Petzoldt has become Pekoldts, Seckgerwitz has been made Sectgerwik, Schultze has been made Schulken, and so on. These obvious mistakes of course have not been ad- mitted. Moreover, several persons are, I suppose, down in the list under two names by which they are known. Thus, Fortius may be Ringelbergius (whose real name was Sterk), Blasius may be Pelacanus. But as I cannot undertake to assert that there is no Fortius except Ringelbergius, &c., I have let both names stand in the list. Again, next to Buckley comes Budæus. Many foreign writers, Heilbronner among the rest, have turned *Buclæus* into *Budæus,* so that in all probability the second of these names is a mistake for the first. And yet there may be also some Budæus who has written on arith- metic ; though, having excluded writers on weights and mea- sures only, I have not put down the author of the work *De Asse.*

I have had, in one or two instances, to throw away German *authors,* for a very obvious reason. The reader will not find the works of *Anleitung,* or *Grundriss,* or *Rechenbuch* in my list, which is more than can be said of every one which has preceded it.

I have not attempted to translate the names of those who

* It is necessary, for instance, to keep close watch upon a writer who introduces among his English authors *Gul. de Cavendy dux de Xeucathle.*

wrote in Latin, at a time when that language was the universal medium of communication. In every such case I consider that the Latin name is that which the author has left to posterity; and that the practice of retaining it is convenient, as marking, to a certain extent, the epoch of his writings, and as being the appellation by which his contemporaries and successors cite him. It is well to know that Copernicus, Dasypodius, Xylander, Regiomontanus, and Clavius were Zepernik, Rauchfuss, Holtzmann, Muller, and Schlüssel. But as the butchers' bills of these eminent men are all lost, and their writings only remain, it is best to designate them by the name which they bear on the latter, rather than the former. In some cases, as in that of Regiomontanus, both names are frequently used: in others, where the Latin consists in a termination only, as Tonstallus for Tonstall, or Paciolus for Pacioli, it matters nothing which is used. It may happen that errors are introduced by returning to the vernacular in a wrong way. I should like to know how it is shewn that Orontius Fineus was Oronce Finée: or in what respect this reading is more finically correct than Horonce Phine, which has great antiquity in its favour? In the case of Vieta, Viète is certainly wrong: he was called by his contemporaries Viet (which I suspect to be derived from the Latin form) and de Viette, never Viète.

It may also be asked, why the unlatinizing process should, for confusion's sake, be practised by the learned only; when it is pretty certain that the world at large will never reconvert Melancthon into Schwartzerd, or Confucius into Kaen-foo-tzee. Neither will they restore to the Popes and other priests of the Roman Catholic Church the names under which they were born and educated.

In many cases it would be impossible to recover these last. For myself, I am well content with the name under which an author was known in literature to his contemporaries, and has been handed down to us, his successors. I know of no canon under which it is imperative to speak of a writer rather as his personal acquaintance than as his reader: and, so far as feeling of congruity is concerned, I think *Alexander ab Alexandro* looks better at the head of a Latin preface than *Saunders Saunderson*. Those who really wish to catch the tone of the middle ages, and shew themselves quite at home, should dwell on the *Christian* name, and make the surname a secondary distinction; should learn to think of *Nicolas* (who happened to be called Copernicus, or Zepernik, it should matter little which) and *Christopher*, to whom the calendar was entrusted.

That this list must be very imperfect I am well aware, for I have been able to add many names to it which I never found in any catalogue. But it will not be useless. It may furnish a reason for preserving any work whatsoever which comes in the way of the reader of this book. If the name be in the list, a book should not be destroyed which has been somewhere catalogued and recognised as a portion of the existing materials for the history of science. But if the name be not in the list, it is obvious that there is some curiosity about a writer whose name is not in the most numerous catalogue of arithmetical authors that has ever been collected in any one place. And it is undeniable that every name must be in one or the other predicament.

I now come to the Catalogue which forms the body of this work. The defects, which every one who has examined the lists that are extant, knows to prevail, arise, in great measure, from the titles and descriptions of books being copied by those who had never seen the books themselves. This is not the worst; for a true copy of a true copy is a true copy : many of these accounts have finally become formal representations of informal titles, which, though not originally intended for more than sufficient indication of the book mentioned, have been used as if they had been full and accurate descriptions. There is a large number of works, not much distinguished in the history of science, each of which has, nevertheless, done its part in its day. The minor points of the same history depend much upon these books, which, being neither of typographical curiosity, nor of literary fame, are gradually finding their way either to the waste-paper warehouse, or the public library. These two depositories are almost equally unfavourable to works of no note assuming their place in the annals of the knowledge to the progress of which they have contributed. Take the library of the British Museum, for instance, valuable and useful and accessible as it is : what chance has a work of being known to be there, merely because it is there ? If it be wanted, it can be asked for ; but to be wanted, it must be known. Nobody can *rummage* the library, except those officially employed there, who will only now and then have leisure to turn their opportunities to account in any independent literary undertaking. And it would perhaps be difficult to make any regulation, under which persons not belonging to the institution might have access to see what is there. Nor will the publication of the catalogue do much towards supplying the defect. Titles of books are but vague indicators of their contents; and a catalogue* of half a million of entries, even if its contents could be guessed at by the titles of the books, is not made to be read through.

It would be something towards a complete collection of mathematical bibliography, if those who have occasion to examine old works, and take pleasure in doing it, would add each his quotum, in the shape of description of such works as he has actually seen, without any attempt to appear more learned than his opportunities have made him. This is what I have done in the descriptive catalogue of works on arithmetic, without selection, or other arrangement than order of date. The only reason for a work being in this list, is that it has come in my way: the only reason for one being out of it is, that it has not come in my way, at least at the time of compilation. Whenever any statement is made which is taken from any other writer, it will be put in brackets [ ], except when the statement itself cites an authority: though I have sometimes put the brackets even in the latter case. The only other mode of proceeding would be, to collect lists from authority, naming the sources of information. But, having found so many errors in these sources, when my opportunities have enabled me to bring them to the test, I did not feel inclined to be the tenth transmitter of inaccurate copies. My mistakes shall be of my own making, and it would not be easy to invent one which should want high precedent for its species.

The description of the books in the Catalogue is uniformly as follows. There are given:

1. The place at which the work was printed, in Italics, and generally in English.

2. The date of the title-page, or colophon, or, when both are wanting, of the preface, in words; but not quite at length. Thus instead of fifteen-hundred-and-eighty-eight will be found fifteen-eighty-eight. I am sure that dates will never be given correctly until this plan is universally adopted: for two reasons. First, the chance of error in printing is very much diminished: particularly the risk of a transposition of figures *at the press*. Secondly, the writer, who is, on the whole, and

---

* When the great catalogue of the Museum is published, those who can give house-room to forty or fifty volumes, and time enough to their examination, may have some of the advantage which they would derive from actual access to the books themselves. And those known to be engaged in research will derive a still larger portion of the same advantage from the readiness with which the officers of the Museum will go out of the usual routine of duty to help them.

one date with another, more to be feared than the printer, has time to be accurate. A glance at four numerals and four strokes with the pen, is too rapid a process for certainty: and those who think they can rely upon themselves to stop *every time,* and look at what they have done, will frequently find reason to wish they had been less confident. But *Incidit in Scyllam,* &c.: I had very nearly announced an edition of Cocker as being unmistakeably the seventeen-hundred-and-twentieth (see page 56).

3. The author's name. When an initial only is given, it is because the author has left no more, either in title or preface.

4. As much of the title as will certainly identify the book. In spelling, initial capitals, &c., I have generally followed the author closely. When there is any defect in this respect, I suppose it will be in those works which I had not any opportunity of re-examining while the sheets were in proof. Imitation of type I have not attempted. To have given the full titles would have swelled my book too much: at certain periods the authors of elementary works were much given to write out descriptive chapters in their title-pages; scores of them would each have filled two or three pages of even the smaller type used in the Catalogue.

5. The form in which the work is printed; a matter which will require some explanation.

A folio, quarto, octavo, duodecimo, or smaller work, is now generally known by its size, though not always. In the folio the sheet of paper makes two leaves or four pages, in the quarto four leaves, in the octavo eight, in the duodecimo twelve, and so on. But even a publisher thinks more of size than of the folding of the sheet when he talks about octavo or quarto; and accordingly, when he folds a sheet of paper into *six* leaves, making what ought to be a *sexto* book, he calls it a duodecimo printed in half sheets, because such printing is always done with half-sized paper, or with half-sheets, so as to give a duodecimo size. From a very early period it has been universal to distinguish the sheets by different letters called *signatures.* In the book now before the reader, which is a half-duodecimo (or what I call a *duodecimo in threes*), the first sheet which follows the prefatory matter, B, has B on the first leaf, and B 2 on the third; which is enough for the folder's purpose. But in former times the signatures were generally carried on through half the sheet, and sometimes through the whole. Again, in modern times, no sheet ever goes into and forms part of another; that is, no leaf of any one .sheet ever lies between two leaves of another. But in the sixteenth century, and even later in Italy, it was common enough to print in *quire-fashion*. Imagine a common copybook, written through straightforward, and the string then cut: and suppose it then separates into four double leaves besides the cover. It would then have sixteen pages, the separate double leaves containing severally pages 1, 2, 15, 16; 3, 4, 13, 14; 5, 6, 11, 12; 7, 8, 9, 10. If a book were printed in this way, it would certainly be a folio, if the four double leaves of any one quire or gathering were each a separate sheet: and if the sheet were the usual size, it would give the common folio size. But if each gathering had the same letter on all its sheets, if the above for instance were marked A_1 on page 1, A_2 on page 3, A_3 on page 5, and A_4 on page 7; the book, when made up, would have all the appearance of a more recent octavo in its signatures. In order to give the size of a book, and at the same time to give the means of identifying the edition by its signatures, I have adopted the plan which gives the following rules :

(a) The words *folio, quarto, octavo, duodecimo, decimo-octavo*, refer entirely to size, as completely as in a modern sale-catalogue, the maker of which never looks at the inside of a book to tell its form. All the very modern distinctions of *imperial, royal, crown, atlas, demy*, &c. &c. I have relinquished to paper-makers and publishers, who alone are able to understand them. But in old books, the reader must expect to see the several sizes, each one of them, smaller than in the modern books. When the work is decidedly small of its name, I have noted it by the word 'small.'

(b) When the single word occurs, without any thing more, the signatures are as in the genuine meaning of the word. Thus, as to signatures, *folio* has two leaves to one letter of signature, *quarto* four leaves, *octavo* eight, *duodecimo* twelve ; and of course double the number of pages.

(c) When the modern word occurs with the addition of *in twos*, or *in threes*, &c. the addition expresses the number of *double leaves* which belong to one letter of signature : and which I believe would be found, if the books were taken to pieces, to be in each quire or gathering. Thus, *folio in ones*, or *quarto in twos*, or *octavo in fours*, or *duodecimo in sixes*, would in each case be unnecessary repetition ; for the first word, when alone, is intended to express the gathering in the third. But *folio in twos* would mean the folio size with two double leaves in one quire, *folio in fours* with four double leaves. Thus, a book of the octavo size, with the quarto signatures, is *octavo in twos:* had it been larger, I should have called it *quarto.*

By this means there is something as to size, and something as to signatures, in every description. But whether any book which I call *octavo in twos,* for instance, really was printed on whole sheets, or on half sheets, that is, really was a small quarto, or a divided octavo, is more than I can in any case undertake to say. All I know is, that with these rules, the reader has two indications in every case, to guide him in determining whether the book he has in his hand be the one I describe or not.

This is no unnecessary excess of description. For so frequently, in the sixteenth and seventeenth centuries, were there issues of the same impression under different titles, that all I have done will in many cases only give a presumption as to whether a book in hand is or is not the edition I have described. Were I to begin this work again, I would in every instance make a reference to some battered letter, or defect of lineation, or something which would be pretty certain not to recur in any real *reprint.* Ordinary errata would not be conclusive: for these might be reprinted for want of perceiving the error.

Rules are given for determining the form of printing by the waterlines of the paper, and by the catchwords. It is supposed that the latter are always at the end of the sheet, and also that the waterlines are perpendicular in folio, octavo, and decimo-octavo books, and horizontal in quarto and duodecimo. But in the first place, a great many old books have catchwords at the bottom of every page, many have none at all; and as to the rule of waterlines, I have found exceptions to every case of it. Pacioli's Euclid, *Venice,* fifteen-nine, *folio in fours,* has horizontal waterlines: the Hypomnemata of Stevinus, an undoubted folio, has thick waterlines *both ways.*

6. In the smaller type are entered such remarks as suggested themselves on the manner or matter of the work, or on any point arising out of it. In some cases these have extended themselves to short dissertations, such as, on the geometrical foot, page 5,—on Sacrobosco's knowledge of the Arabic numerals, page 14,—on the invention of $+$ and $-$, page 19,—on the age of Diophantus, page 47,—on the genuineness of Cocker's Arithmetic, page 56. But for the most part I have contrived to keep within a very moderate compass as to what is said under each work.

The principal point on which I have distinguished one work

from another, is as to the use of the *old* or *new* method of performing division, which, more than any other single point,
decides the character of a work. The letters O, N, or ON,
will tell whether the work uses the old, the new, or both.
Not that the verbal distinction is here very correct; for neither
method is older than the other; and both appear in Pacioli.
A description of the now disused method is given by Dr. Peacock, p. 433.

With regard to the works themselves, I have made no selection, as before noticed. No book that I *have* seen during
the compilation has been held too bad to appear; no book
that I *have not* seen too good to be left out. I have had but
one discretion to exercise, namely, to determine the extent to
which algebra should be considered as arithmetic. In the
earlier day, the distinction was slight: I doubt whether I have
not overrated it; but it is not an easy line to draw.

The history of Arithmetic, as the simple art of computation, has found little notice from the historians of mathematics
in general. They shew themselves deficient in the knowledge
of its progress, and of the connexion of that progress with
the rest of their subject. The writers whom I can name as
having attempted—some more and some less—to supply this
defect, are Wallis, Dechâles, Heilbronner, Scheibel, Kästner,
Leslie, Delambre, Peacock, and Libri. I speak of the progress
of arithmetical writings as works on science, independently of
bibliography properly so called, and biography.

Wallis (p. 44 and Additions) was one of those writers whose
works remain the standard of the erudition of their day. His
algebra, so called, is rather the history, theory, and practice of
both arithmetic and algebra. Its miscellaneous and badly in-
dexed form prevents any from knowing what is in it, except
those who make a study of it, which none of our day will do,
unless they intend to go rather deeply into the history of the
exact sciences. But many and many a page by which the
writer intended to gain the credit of research, will be found to
be a transcript from Wallis. As a connected history, however, it is nothing; and as a bibliography, less. For example,
that Regiomontanus used decimal *fractions*—a very common
story—is the consequence of Wallis's confused method of stating that he introduced the *decimal* instead of the *sexagesimal*
radius into trigonometry; the confusion arising from his not
having a clear knowledge of what had been published of Regiomontanus. And again, we find him, though the editor of
an edition of Oughtred, balancing as to which was written first,

Oughtred's *Clavis*, or Harriot's *Praxis* (which were published
in the same year), apparently ignorant that the great method
as to which the two authors were chiefly to be compared, did
not appear in Oughtred's first edition at all. But for all this,
if Wallis be cautiously watched as to books and dates, his works
are most valuable magazines of historical suggestion. From
them might be collected a much better *scientific* history of
arithmetic than existed in his time, or, indeed, in any time
preceding the publication of Dr. Peacock's article on the subject.

Claude Francis Milliet Dechâles was a Jesuit, who published, in four very large folios, a complete course of mathematics, including architecture, carpentery, fireworks, and all
that was then held to belong to the exact sciences (see page
53). The first volume opens with about a hundred pages
(large folio, double column) *de progressu matheseos*, consisting
entirely of description of books, in order of date. The part
relating to arithmetic fills nine of these pages. The whole is
done with much care, and is, for the mode of describing books
current at the time, very accurate ; and the opinions given on
the books shew that Dechâles had read them. But he is
strongly addicted to the very common mistake of judging the
books according to what ought to have been said of them, if
they had been published in his own time. For example, he
finds that there is not sufficient demonstration in Tonstall ;
which is true, absolutely speaking; but Tonstall is a very
Euclid by the side of his contemporaries.

Heilbronner's Historia, &c. (page 69), though it professes
only to give writers up to the beginning of the sixteenth century, makes a particular exception in favour of arithmetic.
Up to the year 1740, about 170 authors are recorded, a great
many of whom he had not seen. There is also historical dissertation on points of arithmetic. This is a work of great
value to the inquirer : he must not rest upon its statements ;
but he will find more than usual materials for further research.

Scheibel (Additions) may be considered as partly repetition,
partly extension, of Heilbronner. He is one of those bibliographers who collect from various sources the names and dates
of more editions than those who know catalogues will readily
believe in.

Kästner (Additions) falls under the following censure from
Dr. Peacock : "The meagre sketch which Kästner has given
of some insulated works on the subject, generally contrives to
omit almost every particular which is essentially connected

with the history of the progress of the science." This is well
merited, inasmuch as the author chose to call his work a history, instead of a *bibliography*; and as the former, nothing can
be more incomplete. I have almost as good a right to call
my work a history, as Kästner his. But, as a bibliography,
it may be urged in defence, that he gave fuller descriptions of
books than his predecessors. Scheibel, Heilbronner, and Dechâles sink into title-writers (and bad ones) before Kästner.

The late Professor *Leslie* (page 89) was one of those men,
the strength and asperity of whose opinions would make it fair
to deal with them as they dealt with others. In his *Philosophy of Arithmetic* he has entered incidentally into much of
its history. He was, by taste, a searcher of old books; and
various dates, &c. occur, which shew that he had more knowledge of books than can be got from catalogues. A few words
will sometimes shew this to a person who has compared the
books with the accounts of them. But, writing in a popular
manner, he does not give references to his authorities, which
is a serious diminution of the value of his work. Of Leslie, as
an historian on controverted points, one principal thing to be
cautious of is, his almost monomaniac antipathy to every thing
Hindoo—a most unfortunate turn for an arithmetical investigator. Those who inquire into this subject will see what he
is in Colebrooke's hands; those who do not, may compare his
description of the *Lilivati*, "a very poor performance, containing merely a few scanty precepts," with the summary of the
contents of that work in the *Penny Cyclopædia*, article *Viga
Ganita*. Leslie also generalises most fearfully every now and
then. He informs us that it was the *practice throughout Europe*
to reduce the rules of arithmetic to memorial verses, and that
Buckley's *Arithmetica Memorativa* appears at one period to
have *gained possession of the schools and colleges of England*.
Now the truth is, that the verses attributed to Sacrobosco had
never even been printed when Leslie wrote; and Buckley, so
far as is known, was printed only once alone, and two or three
times as an appendix to a work on logic. Dr. Peacock expresses the truth in saying that, *before the invention of printing*,
the practice of *writing* memorial verses was common, as appears by manuscript libraries. It is needless to say that, had
the practice of *using them* been common, the presses of the fifteenth and sixteenth centuries would have given them forth in
great numbers. But I cannot learn that any metrical work
was printed in the fifteenth century, except the *Compotus* of
Anianus, and that only once.

Delambre (page 84) wrote on the history of Greek Arithmetic in such a manner as to set that part of the subject fairly going. I doubt if any one since the time of Wallis, at least of their order of note, has exhibited so much of the union of the *scholar* and the *computer*. Of the obligations under which the history of astronomy lies to him, it is unnecessary to speak here, or any where : no one man was ever so closely connected with that science, with its past, its present, and its future, by history, observations, and tables.

Of my opinion of Dr. *Peacock's* work (page 91) I need hardly say any more, after the pains which I have taken to give reasons on every point in which I differ from its author, and to correct every little error which I have found. This I judge to be the most undoubted compliment which can be paid to any work. Having looked carefully over it, with a great many of the works mentioned in it in my hands, it would be a strong evidence in its favour, were it needed, that I have found no more to set right than is there noted, in matters of dates and circumstances. It is much to be wished that this treatise should be published separately :* those who can obtain it will find that it gives life and spirit to the catalogue of books which forms the main part of this work. Up to this moment it is the only work which can be called a history of arithmetic.

M. *Libri's* history of science in Italy (page 95) is the work of a man who, to the character of a mathematician, adds that of a man who is well versed in literature, and a successful collector of the rarest works in that of his own country. There is much which is interesting in its early history of Italian Arithmetic. Unfortunately, it is not yet finished ; and, I am told, is not likely to be speedily resumed.

The history of most of the sciences resembles a river which sinks underground at a certain part of its course, and emerges again at a distant spot, swelled by certain tributaries, which have joined it in the tunnel. Of arithmetic in parti-

* It was once intended to publish these treatises separately. Nine years ago, the proprietors of the *Encyclopædia Metropolitana* so fully intended to publish separately, that they considered themselves aggrieved because I, who had written the *mathematical* article on probabilities, wrote a *popular* work with that word in the title-page, which they alleged, through their agent, was in effect a republication of the former work, &c. Not being able to get them either to litigation or arbitration, I was obliged to write a pamphlet to prove that the charge was frivolous. The pamphlet is unanswered, and all the treatises unpublished (separately) to this day. The latter I regret, for the sake of science. It is a great pity that Sir John Herschel's treatises on light and sound, Dr. Peacock's arithmetic, Mr. Airy's tides, &c. are thus locked up.

cular, we note the disappearance in the seventh century of all that was good in the Greek system; and we see the rise, at the invention of printing, of the most trivial part of it, combined with the additions which are now well known to have been received from Eastern sources.

The manuscript literature of the middle ages will prove no very productive source of information, to judge by all that has been made of it hitherto. But then it is to be remembered how seldom, if ever, it has happened, that the investigator of it has united the character of a sufficient mathematician with that of an industrious and well-trained palæographist. Neither the one nor the other can proceed alone : the former has not only to learn how to read, but what to read; for the actual habitat of the manuscripts, and how to get preliminary knowledge of their existence, is a study of itself. The latter is apt to know no difference between what is sound, and what is worthless.

When Euclid, Ptolemy, &c. are first seen to reappear, they come in as Arabic writers. They may be called Greek, but the first translations are from the Arabic; and their effect upon the literature of Europe is, in the first instance, just what it might have been if the authors had been Persians or Saracens. How the communication took place has been considered as a point of small curiosity compared with the importation into Europe of the Arabic numerals.

This subject, both generally, and with reference to the several countries, has been long on the anvil : not the authorship of the letters of Junius has tasked research and ingenuity more than the introduction of the nine digits and the cipher. I suppose nobody would listen to the hypothesis that the former wrote themselves : but I am much inclined to suspect that the latter introduced themselves. The *endosmose* constantly going on between nations connected by war and commerce would not merely explain so easy a matter, but would render it very difficult to explain how it did not happen, if it had not happened. That the Venetian merchants should not know the system of accounts of those with whom they traded, is incredible : that, at the beginning of the thirteenth century, many priests and some soldiers, returned from their crusades, should not bring back with them the account of so very elementary a difference between themselves and those with whom they had treated, and whom they had in many instances served as slaves taken in war, is unlikely. That Leonard of Pisa, as he is called, was the first who wrote on or in the new system,

is pretty generally affirmed by the Italians of the fifteenth
and sixteenth centuries : and there can be no doubt of it. But
because we trace the first formal user or expounder to Italy,
it by no means follows that the other parts of Europe owe the
system to that country in the first instance, however certain
it may be that they owe much assistance in the course of the
general establishment. It was so common in England in the
thirteenth century, that Roger Bacon recommends it (page 14)
as a study, of course as a practicable study, for the clergy.
Had it not been commonly known, at least as to what it *was*,
he would have done more than give the name of the science,
and the names of its rules ; he would have added some description. Those who have searched into the matter, merely with
a view to Arabic *numerals*, and without any collateral thoughts
about *arithmetic* in their heads, may have passed over much
valuable evidence. They did not know, perhaps, that the
organised rules of computation always went with the Arabic
system, and never with the Roman or Boethian : that if ever
they came to the connected mention of addition, subtraction,
multiplication, and division, it ought to have been a sign that
they were reading on *algorism* as distinguished from *arithmetic*. This passage of Roger Bacon has been neglected ;
though the mere occurrence of the word *algorism*, in a most
interpolable clause of one sentence, occurring in Matthew Paris,
has received notice and discussion from Mr. Hallam, from a
learned writer in the Archæologia, and probably from others.

The rejection of the work attributed to Sacrobosco may be
accounted for from the general ignorance of its having been
printed under his name at an early period. Though this does
not prove the genuineness of the work, it very much adds to
the evidence for it. As early as 1523, the learned world was
invited to dissent from the assertion that the treatise in question was written by Sacrobosco, and did not do it. It was
often cited afterwards, and never, as far as I have seen, with
any doubt. But the only writer of this or the last century
that I can find, who describes the *printed* edition, is the Italian editor of Fabricius, *Bibliotheca Latina*.

Come how they might, however, we find ourselves, at the
invention of printing, in possession of the knowledge that two
distinct systems of arithmetic were current. The streams
which had united in one bed had not mingled their waters ;
and for nearly a century two distinct classes of writings present themselves. Their only common point is the use of the
Arabic or Indian numerals and method of notation.

The first, or *algoristic* system, as we may call it, proceeded by systematic rules to the performance of questions of computation. It embraced mercantile arithmetic, and what was then called algebra: and laborious and prolix efforts were made to combine the two; that is, to represent in such a form as we should now call algebraic, as distinguished from arithmetical, the solutions of questions of commerce, or of what might become such, if the connexion were diligently fostered. I cannot suppose that the multifarious problems of exchange of different kinds which abound in Pacioli and his immediate followers, were actually useful to the merchant: but there is an obvious leaning to the idea that they ought to be so. Geometry was also an application of this arithmetic; and in a manner which strongly marks its eastern character, as will appear to those who compare the work of Pacioli with the Indian books.

The second, or *Boethian* system, as I shall call it, because the work of Boethius was its great text-book, did not give any rules of calculation, nor apply itself to any application. The study was the properties of numbers, and particularly of their ratios. There was no art about it; and we have no means of telling whether the philosophers of this school reckoned on their fingers, or used an abacus, or put pen to paper for the performance of some organised method of computation. To judge by the smallness of the numbers used in the instances adduced, we must suppose that the writers left it to their readers to do as they liked best.

For some specimens of the laborious manner by which the Pythagorean Greeks, in the first instance, and afterwards Boethius in Latin, had endeavoured to systematise the expression of numerical ratios, I may refer the reader to the article "Numbers, old appellations of," in the Supplement to the *Penny Cyclopædia*. If I were to give any account of the whole system, on a scale commensurate with the magnitude of the works written on it, the reader's patience would not be *subquatuordecupla subsuperbipartiens septimas* — or, as we should now say, *seven per cent*—of what he would find wanted for the occasion. Not that the books I am speaking of get quite so far as this. I am hardly prepared to say exactly what number under fifty ought to be named as the terminus of the Boethian store of numbers; but certainly we rarely find them choose their instances from numbers above it. The system broke down under its own phraseology, as did that of the Yancos mentioned by Dr. Peacock, who could not get further than *three*, because they could not express this idea by any thing more simple than *Poettarrarorincoaroac*.

The Italian school of *algorists*, with Pacioli at their head, found followers in Germany, England, France, and Spain ; and in all but England, the Boethian school also. I cannot discover a single English work which pays any detailed attention to the latter class of arithmeticians. Tonstall, and still more Recorde, for the former appears to have been little known, were the preservers; and before the end of the sixteenth century the ordinary style of commercial arithmetic, which has prevailed among us ever since, was in course of establishment.

This gradual formation of the English school of commercial works will be apparent enough in my list. From the time of Recorde, we always were conspicuous in numerical skill as applied to money. The questions of the English books are harder, involve more figures in the data, and are more skilfully solved. It is possible that I might, if my French, German, and Italian list were more complete, produce exceptions to this rule. But nothing, I think, could arise to alter my conviction that the efforts which were made in this country towards the completion of the logarithmic tables in the seventeenth century, and the instantaneous appreciation of the value of the discovery of logarithms, were the result of that superiority in calculation which I assert to have been formed in the sixteenth. And yet this last-named century produced one man in France, one in Germany, and one in Italy, with either of whom no one English calculator could compare in extent of operations : I allude to Vieta, Rheticus, and Cataldi. There was no opportunity to compete with these men ; for the subjects on which they worked were not introduced here. It was only towards the *end* of the sixteenth century that what were then the higher parts of the mathematical sciences began to be disseminated with effect in Britain.

To the commercial school of arithmeticians above noted we owe the destruction of demonstrative arithmetic in this country, or rather the prevention of its growth. It never was much the habit of arithmeticians to prove their rules : and the very word *proof*, in that science, never came to mean more* than a test of the correctness of a particular operation, by reversing the process, casting out the nines, or the like. As

* At first I had written "degenerated into nothing more;" but this is incorrect. The original meaning of the word *proof*, in our language, is testing by trial.

c 2

soon as attention was fairly diverted to arithmetic for commercial purposes alone, such rational explanation as had been handed down from the writers of the sixteenth century began to disappear, and was finally extinct in the work of Cocker, or Hawkins, as I think I have shewn reason for supposing it should be called. From this time began the finished school of teachers, whose pupils ask, when a question is given, what rule it is in, and run away, when they grow up, from any numerical statement, with the declaration that any thing may be proved by figures—as it may, to them. Any thing may be unanswerably propounded, by means of figures, to those who cannot think upon number. Towards the end of the last century, we see a succession of works, arising one after the other, all complaining of the state into which arithmetic had fallen, all professing to give rational explanation, and hardly one making a single step in advance of its predecessors.

It may very well be doubted whether the earlier arithmeticians could have given general demonstrations of their processes. It is an unquestionable fact of observation, that the application of elementary principles to their apparently most natural deductions, without drawing upon subsequent, or what ought to be subsequent, combinations, seldom takes place at the commencement of any branch of science. It is the work of advanced thought. But the earlier arithmeticians and algebraists had another difficulty to contend with : their fear of their own half-understood conclusions, and the caution with which it obliged them to proceed in extending their half-formed language. It was not merely by oversight, I suspect, that Oughtred so often calculates $ab + ac$ by two multiplications, instead of using $a(b+c)$: but rather from that same general fear of abbreviation, and suspicion that error may lurk in it, which possesses men of business who dare not multiply by 10 by the annexation of a cipher, but proceed with each figure, and carriage, as they would do if the multiplier were 8 or 7. I have seen this often enough ; and things nearly as strange times without number. But what shall we say to the following ; a most sufficient recommendation of the study of old works to the teacher, as shewing that the difficulties which it is now (I speak to the *teacher* not the *rule-driller*) his business to make smooth to the youngest learners, are precisely those which formerly stood in the way of the greatest minds, and sometimes effectually stopped their progress. Perhaps no man of his day had so much power over mathematical language as Wallis. But the following extract of a letter from

him to Collins (Macclesfield Collection, vol. ii. p. 579), in 1673,
shews that he once had doubts whether he might dare write
down the square root of 12 as being twice the square root of 3,
however certain he might be that it is so; because no one
had so written it. Speaking of the square root of a negative
quantity, he says, "Only, though I had from the first a good
mind to it, I durst not without a precedent, when I was so
young an algebraist as in the history my late letter reports,
take upon me to introduce a new way of notation, which I
did not know of any to have used before me. And it was not
without some diffidence that I ventured on $2\sqrt{3}$, instead of
$\sqrt{12}$, not having then met with any example of a number so
prefixed to a surd root; but I found it so expedient, not only
for the discovering the root of a binomial, whether quadratic,
cubic, or others, but for the adding and subducting of com-
mensurable surds, that I resolved to use it for my own occasions, before I knew whether others would approve of it or
no; especially having found in the first edition of Oughtred's
Clavis (for I had not then seen the second) one instance or
two for it, to justify myself if it should be questioned. But
since that time it is grown more common, and I perhaps have
somewhat contributed thereunto."

If I were to take such things from old and unknown
writers, in which they abound, I should be thought to waste
the time of my readers by giving an undue importance to the
difficulties of very inferior minds. But those who do not want
such a confession as the above from a Wallis, to help to per-
suade them that the difficulties of progress are of the same
character in all ages, will find in old works plentiful and useful supplies of thought for a modern teacher.

Another instance of the halting progress of language is
the mode of introduction of the decimal point. Stevinus, the
undoubted introducer of the decimal fraction (though others
may have seen very clearly the great use of 10, 100, 1000, &c.
as divisors), had no such thing, but only a distinct mode of de-
noting the meaning of each decimal place. Dr. Peacock mentions Napier as being the person to whom the introduction is
unquestionably due: a position which I must dispute, upon
additional evidence.

The inventor of the single decimal distinction, be it point
or line, as in 123·456 or 123|456, is the person who first
made this distinction a permanent language: not using it
merely as a rest in a process, to be useful in pointing out afterwards how another process is to come on, or language is to

be applied; but making it his final and permanent indication
as well of the way of pointing out where the integers end and
the fractions begin, as of the manner in which that distinction
modifies operations. Now, first, I submit that Napier did not
do this; secondly, that if he did do it, Richard Witt did it
before him (page 34).

I have not seen Wright's summary of logarithms, of 1616,
to which Dr. Peacock refers for some indications of the decimal point. But I take with confidence his assertion that the
first distinct notice of the thing in question, by or by means
of Napier, is contained in the *Rabdologia* (page 35).

In this famous tract there are but two instances to the purpose: a scanty number, be it noticed, for a person who had
seized the idea of completing the decimal scale. The first is
an instance of division which Dr. Peacock cites. Here the
quotient is, as we should now say, 1993·273. Napier uses a
comma in his quotient, as a rest, and writes 1993,273, and
then presents his answer in the form 19932′7″3‴, as Stevinus
would have done. Dr. Peacock states this, and notes it as a
"partial conformity to the practice of Stevinus." Unfortunately Napier has not given us instances enough to tell us whether he had a *practice* of his own: and in the only other case,
which Dr. Peacock also cites (I have verified both citations),
he adheres still more closely to the method of Stevinus. For
though, in a question of addition, he has drawn a line down
the interval where our decimal points would be, which line
would be as distinctive in his total as in his *addenda*, he places
the exponents of Stevinus in this total, which thus has both
distinctions, and stands 1994|9′1″6‴0⁗.

I cannot trace the decimal point in this: but if required
to do so, I can see it more distinctly in Witt (page 33), who
published four years before Napier. But I can hardly admit
him to have arrived at the notation of the decimal point. For
though his tables are most distinctly stated to contain only
numerators, the denominator of which is always unity followed
by ciphers; and though he has arrived at a complete and permanent command of the decimal *separator* (which with him
is a vertical line) in every operation, as is proved by many
scores of instances; and though he never thinks of multiplying or dividing by a power of 10 in any other way than by
altering the place of this decimal separator; yet I cannot see
any reason to suppose that he gave a meaning to the quantity
with its separator inserted. I apprehend that if asked what
his 123|456 was, he would have answered: It *gives* $123\frac{456}{1000}$,

not it *is* $123\frac{456}{1000}$. This is a wire-drawn distinction : but what
mathematician is there who does not know the great difference
which so slight a change of idea has often led to? The person
who first distinctly saw that the answer -7 always implies
that the problem requires 7 things of the kind diametrically
opposed to those which were assumed in the reasoning, made
a great step in algebra. But some other stepped over his
head, who first proposed to let -7 *stand for* 7 such diametrically opposite things.

Who, then, was the real inventor of the decimal separator,
to give both interpretation and operative form? This is a
question I will not answer positively : nor attempt to answer
till I have pointed out how the case stands with several possible claimants. And first, as to Briggs. He does not indeed
arrive at the simple decimal *point* (which is a strong presumption against such a thing being suggested by Napier ; for who
would have learnt it from him, if not Briggs?)—but he omits
the denominator, and draws a line under the decimals. And
it is a further presumption against any such idea of abbreviation being common, that Briggs explains his notation as quite
peculiar to himself. In the preface, or *Ad Lectorem*, of the
Arithmetica Logarithmica, London, sixteen-twenty-four, *folio in
twos*, he commences thus:

"Ut obscuritatis crimen, quantum in me situm fuerit
effugiam, paucula hæc humanissime Lector, te admonendum
censui. 1. Si numerus occurrat cui linea subscribitur, notæ
illæ quæ supra lineam sunt descriptæ, Numeratorem partium
constituunt, quarum Denominator semper intelligitur unitas,
cum tot cyphris, quot sunt notæ superius positæ. ut 75 de-
signant $\frac{75}{100}$ vel $\frac{3}{4}$. sic 5 9321 designant $5\frac{9321}{10000}$."

Oughtred adopted both the vertical and sub-horizontal
separators, thus shutting up the numerator in a semi-rectan-
gular outline. In the mean time, Gunter, after adopting
Briggs's notation in the first instance, gradually dropt it, and
substituted the decimal point. The history of this symbol in
his hands is rather curious, on account of its having been so
completely overlooked, though all the circumstances were to
be seen in widely circulated works.

Gunter published in 1620 his *Canon Logarithmorum*, sines
and tangents on Briggs's system. To this table there is no
explanation. But an explanation was written by him, and,
though I cannot make out that it was ever published separately, yet it appears in the collection of his works. I have
what is called the second edition of this collection, *London,*

sixteen-thirty-six, *quarto*, which, for various reasons, I suspect
to be the first, thinking that the announcement of second edi-
tiou merely refers to the fact of its having been previously
published during Gunter's life. After the *finis*, with new
paging and signatures, comes 'The Generall use of the Canon
and Table of Logarithmes.' This I suspect to have then made
its first appearance: but from the very manner in which the
decimal fractions are treated, it is clear that it was written
before or with the work on the sector, &c., which first* came
out in sixteen-twenty-three; and this, though it makes reference to that work, or at least to its matter. At the beginning
of this 'Generall use, &c.' common fractions are used, even
when the denominators are decimal. At page 13, Briggs's notation appears, without explanation: and 116 04 is the third
proportional to 100 and 108; this continues through page 14.
In page 15, a dot is added to Briggs's notation in one instance:
100*l*. in 20 years at 8 per cent becomes 466.095*l*. At the
bottom of this page Briggs's notation disappears thus, "It appeareth before, that 100*l*. due at the yeares end is worth but
92 592 in ready money: If it be due at the end of 2 yeares,
the present worth is 85*l*. 733: then adding these two together,
wee have 178*l*. 326 for the present worth of 100. pound an-
nuity for 2 yeeres and so forward." After this change, thus
made without warning† in the middle of a sentence, Briggs'
notation occurs no more in the part which relates to numbers.
But in the following chapter, which is trigonometrical (and
which may have been written first, for Gunter's *own* logarithms are only trigonometrical), it reappears, sometimes with
and sometimes without the dot. In the previous work on the
sector, &c., the simple point is always *used*: but in explanation the fraction is not thus written, but described as parts.
Thus, 32.81 feet in operation is 32 feet 81 parts in the description of the question or answer.

It was long before the simple decimal point was fully
recognised in all its uses, in England at least: and on the
continent the writers were rather behind ours in this matter.
As long as Oughtred was widely used, that is, till the end of
the seventeenth century, there must always have been a large
school of those who were trained to the notation 123|456. To

---

* A printed title-page has 1623: an additional engraved and ornamented title-page has 1624.
† As far as I can find, the words *cosine* and *cotangent* (which Gunter
introduced) appear in the same manner, without warning or explanation.

the first quarter of the eighteenth century, then, we must refer, not only the complete and final victory of the decimal point, but also that of the now universal method of performing the operation of division and extraction of the square root.

This evident slowness in the admission of such important improvements will not appear singular to those who observe what is now taking place with reference to Horner's method for the solution of equations. It will be the business of some successor of mine, a century or two hence, to distinguish in his list, by some appropriate mark, the works of our period which adopt this method from those which do not: just as I have done with the two methods of division. I hope, if I live to publish a second edition, that I may be able to add to my list the name of the *first* work, written for students in the University of Cambridge, which shall contain this most fundamental addition to the processes of pure arithmetic : which, though it has found its way into the public examinations, is not yet in the books which prepare students for them.

This has been a long digression, on what is perhaps the most interesting part of the subject, its language. We owe every thing, almost, to the simplicity of certain modes of expression. Nothing is more clear than that the Greek geometers, with all the acuteness and perseverance necessary to carry results of arithmetical application to a high pitch, and all the taste for numbers which would turn their thoughts that way, were stopped by their insufficient system of numeration, and the tediousness of their processes. The history of language, then, is of the highest order of interest, as well as utility : its suggestions are the best lessons for the future which a reflecting mind can have.

In looking over the list of books, it will be observed, that during the eighteenth and nineteenth centuries there is a great preponderance of English writers. The law I imposed on myself, of entering no books which I could not see, necessarily brought this about. During the last two centuries, the elementary works of the different countries in Europe have not gained great general circulation. They have ceased to be written by the greatest names, for the most part; and they have rarely been cited by the historians of science. Copies which have found their way to this country have probably soon been destroyed. As it is my intention to endeavour to extend this list, whether I publish the extensions or not, and in any case to provide for the preservation of the materials I collect, it will be worth the while of any one who is able, to

furnish me with information on works which I have not seen.
I think it probable that any one who has had the curiosity to
rescue three books of arithmetic from a stall, will find that
one of them is not in this list.   From any such person I shall
thankfully receive the full title of the neglected work, with
the form in which it is printed.   Nor need it by any means
be presumed that because a book is wholly unknown, it proves
nothing in the history of the science.   A book so thoroughly
lost as that of Witt, contains a nearer approach to the decimal
point than was made by Napier.   Horner found a close approximation to his own method, for the case of the cube root,
in an obscure compendium of arithmetic.   I might also instance
Dary's discovery mentioned in page 48 ; and other things of
the same kind.   The history of hints given before the time at
which they were (perhaps could be) made to bear fruit, would
be a very curious one : and the progress of science will never
be well understood until some little account can, in each case,
be given of the reason why a notion should be so productive
at a particular period, which was so barren at a previous one.

On looking at the fourth edition of Gunter's work (sixteen-
sixty-two), I find that some liberties have been taken with the original text : among others, Briggs's notation is restored in several
places, though not entirely.   It should also be noticed that the
point which generally occurs after the first figure of a logarithm
is not a fractional separator, but only divides one integer from the
rest.

A LIST

WORKS ON ARITHMETIC,

THEORETICAL AND COMMERCIAL, IN CHRONOLOGICAL ORDER.

** The letter P. followed by a number, refers to the page of Dr. Peacock's treatise in which the work or its author is mentioned. The brackets [ ] enclose statements which I have taken from others, and cannot therefore affirm from personal inspection. For O and N see those letters in the Index.

Florence, fourteen-ninety-one. **Philip Calandri.** 'Philippi Calandri ad nobilem et studiosum Julianum Laurentii Medicem de Arimethrica opusculum.' *Octavo* (small).

This book, which very few have mentioned at all, and fewer still from inspection, is a part of the rich bequest of the late Mr. Grenville to the nation. It begins with a picture of Pythagoras teaching, headed 'Pictagoras Arithmetrice introductor.' In the preface is the following on Leonard of Pisa: 'Vero e che il modo del notare e numeri con decte figure dice Lionardo pisano haver nel Mcc. incirca rechato dindia in Italia : et decti carateri: o vero figure essere indiane : et appresso deglindi havere imparato la copulatione desse.' He explains the leading rules, except division, for integers and for lire, soldi, and denari. Division he puts down in examples, and appears to have mistaken the mode of working, or to have had an incorrect printer. His notion of a divisor is curious. When he divides by 8, he calls his divisor 7; demanding, as it were, that quotient which, with *seven* more like itself, will make the dividend. He also describes the rules for fractions, and gives some geometrical and other applications. The book is in black letter, but the numerals are in a thin and small-bodied type. At the end of the book is, 'Impresso nella excelsa cipta de Firenze per Lorenzo de Morgiani et Giovanni Thedesco da Maganza finito a di primo di Bennaio 1491.'

B

2   A CHRONOLOGICAL LIST OF

Venice, M.cccc.lxliiii., fourteen-ninety-four. **Lucas Pacioli,** de Burgo Sancti Sepulcri. 'Summa de Arithmetica
Geometria Proportioni et Proportionalita.' *Folio in fours.*
Tusculano, fifteen-twenty-three. **Lucas Pacioli.** 'Summa de Arithmetica geometria. Proportioni: et proportionalita :
Novamente impressa In Toscolano su la riva dil Benacense
et unico carpionista Laco : Amenissimo Sito : de li antique
et evidenti ruine di la nobil cita Benaco ditta illustrato : Cum
numerosita de Imperatorii epithaphii di antique et perfette
littere sculpiti dotato : et continens* finissimi et mirabil colone
marmorei : inumeri fragmenti di alabastro porphidi et ser-
pentini. Cose certo lettor mio diletto oculata fide miratu digne
sotterra se ritrovano.' *Folio in fours* (ON).

These are the full titles of the two editions of this celebrated
work. Both editions are by one printer, Paganino of Brescia, and
are beautifully printed : the type of the second being a good instance
of the black letter in its state of approach to what is now called Roman letter. The work itself has been described by Hutton, Montucla, Peacock, Libri, &c.; but it would yet require a volume of
description to do it justice. It is sometimes called the first work on
arithmetic printed ; but Calandri, Peter Borgo, three already mentioned in the Introduction, and perhaps more, take precedence. But
it is certainly the first printed on algebra, and probably the first on
book-keeping. P. 414, 424, 429 &c., 432 &c., 451, 460, 462 ; Hutton, *Tracts,* vol. ii. p. 201.

On comparing my copy of the first edition with that in the British
Museum, I found one of those phenomena which so frequently
occur in very old printed books. The first leaves of the two copies,
to the number of about thirty, and the first leaf of the geometry,
are not from the same setting up. The endings of the pages, and
the ornaments of the capital letters, are different in the two. Nor
does either of them agree with the second edition. A part of the
first impression may have been lost, so that a second setting of
types was required to replace that part. Several mathematical bibliographers of note enter Pacioli as Lucas di Borgo, and even as Bor-
godi, Lucas, and some do not mention the first edition.

The Latin *Paciolus* is usually spelt in Italian Paccioli, but Libri
spells it Pacioli. I believe that the various assertions that Pacioli
wrote Euclid in *Italian,* arise from the geometry of the work above
described being Italian : but it is not Euclid, though of course
founded on him.

There is another old book of which I have reserved the mention till now, because its author has been confounded with Pacioli.
Brunet gives it as 'La Nobel opera di Arithmetica . . . compilata
per B. [should be P.] Borgi,' *Venice,* fourteen-eighty-four, *quarto :*
and Hain has ' Borgo (Pietro) Venet. Aritmetica,' *Venice,* fourteen-

* Cns in the title.

eighty-two, *quarto*. This edition Hain had not seen: but he had seen the next (the one mentioned by Brunet), which begins 'Qui comenza la nobel opera di arithmetica . . . per Piero Borgi,' *Venice*, fourteen-eighty-four, *quarto*. And he gives two other editions, *Venice*, fourteen-eighty-eight and ninety-one, *quarto*. Accordingly, Mattaire sets down the first edition of *Lucas di Borgo* as of 1484, and Dr. Peacock adopts this date. It is quite certain that Dr. Peacock must have had one of the editions of Pacioli before him when he wrote ; and we are therefore to suppose that, in stating the date of the first edition, he followed Mattaire, and presumed an edition earlier than any he had seen. Tartaglia refers to Peter Borgo. P. 458.

No place, no date. **John Muris.** 'Arithmetices Compendium ex Boetii Libris per Johannem Muris excellentis ingenii virum Accurate congestum.' *Quarto*.

This is a different edition from the one presently mentioned. John Muris seems to have escaped the notice of bibliographers as an arithmetician. I think I remember that there is in Hawkins's History of Music some discussion of his ideas on the musical scale, [on which was published his work, *Leipsic*, fourteen-ninety-six, *folio*]. Muris lived before the invention of printing, but when I cannot ascertain. [His astronomical tables are preserved in manuscript.] The present tract has twelve leaves unpaged, and is certainly of the very earliest part of the sixteenth century, if not of the fifteenth.

Venice, fifteen-one. **Geo. Valla.** 'De expetendis et fugiendis rebus opus.' *Folio in fours*.

The first part of this is ' De Arithmetica libri iii, ubi quædam a Boetio pretermissa tractantur' (O). Certainly he supplies some of the omissions of Boethius; the four rules for instance. The work was published by the son, Joh. Pet. Valla. I have seen an earlier edition mentioned, but I cannot find any trustworthy account of such a thing.

Cologne, fifteen-one. **John Huswirt.** 'Enchiridion Novus Algorismi summopere visus De integris. Minutiis vulgaribus Projectilibus Et regulis mercatorum sine figurarum (more ytalorum) deletione percommode tractans,' &c. *Quarto*, *twos and threes* (O).

A short treatise, apparently one of the earliest printed in Germany on the Arabic system. The *projectilia* are counters. The rules are verified by casting out the nines. The words ' sine figurarum deletione' are not made good in the rule of division. On this work see the *Companion to the Almanac* for 1844, p. 3.

Paris, fifteen-three. **Joh. Faber Stapulensis, Jodocus Chlichtoveus,** and **Carolus Bovillus.** ' In hoc libro contenta Epitome compendiosaque introductio in libros Arithmeticos divi Severini Boethii,' &c. &c.   *Folio in fours.*

This book contains an epitome of Boethius by Faber of Étaples; with a commentary and an arithmetical collection of rules by Chlichtoveus, his pupil; a compendium of geometry, a book on the quadrature of the circle, and *cubication* of the sphere, and a book on perspective, by Charles Bovillus; with an astronomical compendium by Faber already mentioned. This book is one of the earliest printed by Henry Stephens and his partner Wolfgang Hopilius. It is the first edition, I have no doubt, of Faber's epitome of Boethius, though Heilbronner and Murhard assert the contrary, perhaps misled by Faber's edition of Jordanus. As to the contents, the Arithmetic of Boethius was the classical work of the middle ages. It consists of statements of the commonest properties of numbers, under a great many classifications, to each of which a name is given. The second work, of arithmetical rules, shews the very low state of the art. It takes pages upon pages to explain the simple rules, though no examples are ventured on which have more than three figures. Montucla says that the quadrature of Bovillus was only saved from the laughter of geometers by its obscurity. But a work printed by Henry Stephens, and containing Boetius and Faber, must have been very far from obscure in its day. The historians of mathematics confine themselves to works the reputation of which has lasted: but they ought not to make the state of their own minds, with respect to the rest, a criterion of that of the contemporary readers.

Basle, fifteen-eight.   **Gregorius Reisch.**   'Margarita Philosophica, cum additionibus novis: ab auctore suo studiosissima revisione tertio superadditis.'   *Quarto in fours.*

According to Kloss (from his own copies) the first edition of this curious book is *Friberg* fifteen-three, the second and third both *Strasburg* fifteen-four (one printed by Gruninger, which I have seen, and one by Schott), and he calls the one before me the fourth. But Hain marks as the first edition one which appears to be printed at *Heidelberg,* fourteen-ninety-six, *quarto.* The one before me is not the third; it is evident that a former title-page has been reprinted: for though this edition is printed (as appears from the colophon) by Furter and Scotus of Basle, the title-page bears ' Jo. Schottus Argent. lectori S.' Which agrees with Kloss's statement that the previous edition to this one was printed by Schott of Strasburg. The Arithmetic (O), which is a part of the system of philosophy here laid down, has a frontispiece representing Boethius at one table with Arabic numerals before him; and Pythagoras at another with counters. Pythagoras among the Greeks, Apuleius and Boethius among the Romans, were rather made the inventors of arithmetic. The arithmetic is divided into speculative and practical. The former is a summary of Boethius, often in the words of John de Muris. The latter is a short treatise on *Algorithm,* as it was called, or the rules of computation by the Arabic numerals. There is also computation by counters, fractions common and sexagesimal, and the rule of three. Many works of fifty years later do no more difficult questions. That Boethius was the author of the Arabic numerals was a common notion at the time, revived in our own day. I have seen another edition of *Strasburg*, fifteen-twelve; and there is said to be another edition by Orontius Fineus, *Paris* (?) fifteen-twenty. There is also an Italian translation, *Venice*, fifteen-ninety-nine, *quarto*, with the additions of Orontius, translated by Giovan Paolo Galucci.

If the number were sufficient of those who wish to take their notions of liberal education in Europe at the time immediately preceding the Reformation from original sources, and not from the reports of others, a reprint of the *Margarita Philosophica* would be made. The diversity of the matters which it treats, and the largeness of its circulation, stamp it as the best book for such a purpose.

The *Margarita Philosophica* is the earliest work I have found in which mention is made of that peculiar system of measures which was current among the mathematicians of the sixteenth century, and which has caused no little confusion among writers on metrology. I have already given some account of this ill-understood system, and shall here endeavour to present the whole case with some additional evidence.

The Roman foot (of 11·62 inches English) was of course established throughout the empire : and with it the Roman pace* of five feet, or 58·1 inches or 4·84 feet English. The natural pace of a man in our day is as nearly as possible five English feet : Paucton's experiments on the walking paces of individuals gave him 59·7 inches English, or 4·98 feet. In the British army the step, both what is called the ordinary step and the quick step, is, by regulation, thirty inches : making a pace of five feet. The Roman pace, by which distances were actually measured, was that of a soldier on the march : and, as might be expected, the weight of his arms and other equipage seems to have shortened his pace a little. But so near were these measures, as actually used by the Romans, to the natural ones from which they derived their names, that it was customary, not only to recur to legal standards, but, in the absence of ready access to them, to make use of the natural foot and pace. And we also know from Roman writers, that a somewhat fanciful relation, though not very far from the truth, was established between the breadth of the hand across the middle of the fingers, the *palm*, and the length of the *foot*. It was taken that four palms made a foot : and a palm was made of course to consist of four (average) finger-*breadths* or digits. This division into palms and digits was the most recognised division of the Roman foot : that into inches, or *unciæ*, is well known not to belong to the foot merely, but to any thing else. Whatever magnitude was called unity, the *uncia* was its twelfth part. Accordingly, *all* the Roman foot-rules which have been found in ruins or excava-

* The pace is derived from the double step, being the distance from the extremity of the heel at the place from which it is removed in walking to the same at the place in which it is set down again.

tions have the digital division, to which *some* (most, I believe) have the uncial division superadded. All this I take to be too well established to require the citation of any authorities.

It is quite out of my power, or, as far as I know, of that of any one else, to trace the gradual alteration of the foot in different countries. It does not appear that any means were taken to institute comparisons of various measures, to be preserved as public records. Such means could have been found: that which was then the common church of Christendom might have easily regulated the weights and measures of Europe, even without appearing to do so. But in all probability, the extent of the variations was not well known until it was too late. We know, however, as a fact, that the geometers were successful in establishing a measure among themselves, and communicating it through Europe on paper. This measure, I have no doubt, they believed to be the true Roman foot: for they divide it in Roman denominations, make use of it in their quotations from Roman authors, and never hint at their having any other notion of a Roman foot. And moreover, the writers who, in the sixteenth century, recovered the true Roman foot, never mention any peculiar geometrical foot in use among mathematicians, or in any way distinguish the latter from the *wrong* Roman foot which they were correcting. That the geometers believed the Roman foot, that is, their own foot, to be the *human* foot, might be easily proved. And, with such belief, they would make their so-called Roman foot too short. From a hundred measures of the feet of adult men, furnished to me by a boot-maker, and taken as they came in his books, I find the average length of the Englishman's foot to be 10·26 inches, in our day: or an inch and a third shorter than the Roman lineal measure.

This *geometrical foot* of the mathematicians is, I make no doubt, the geometrical foot to which writers of the seventeenth century refer, or mean to refer. But, not long after the true restoration of the ancient measures, there arose a disposition among those who inquired into the subject to seek a mystical origin of weights and measures, on the supposition of some body of exact science once existing, but now only seen in its vestiges: a disposition which is not yet entirely extinct. Some speculated on the pyramids of Egypt, and tried to establish that the intention of building those great masses was that a record of measures founded on the most exact principles might exist for ever. But more turned their attention to the measurement of the earth, and, by assuming nothing more difficult than that a degree of the meridian a thousand times more accurate than that of Eratosthenes was in existence hundreds, if not thousands, of years before him, it was easy enough to make out that the whole system of Greek, Roman, Asiatic, Egyptian, &c. measures was a tradition from, or a corruption of, this venerable piece of lost geodesy. There runs through all these national systems a certain resemblance in the measures of length : if a bundle of faggots were made of foot-rules, one from every nation ancient and modern, there would not be any very unreasonable difference in the lengths of the sticks.

The metrologists who treat this subject handle it according to their several theories. Those who have none in particular either neglect it altogether, or speak of its length as uncertain, or define, with Dr. Bernard, the geometrical pace as being five feet of its own kind, without saying what this kind is. Those who have the notion of the old measure of the meridian accommodate it to their supposed ancient measure; but at the same time, those of most research and note make it *less* than their Roman foot. Thus Paucton makes it nine-tenths of the Roman foot, which, with his version of that measure, is 10·9 inches English, and with the true one, 10·5 of the same. Similarly, Romé de L'Isle makes it more than half an inch (French) less than *his* Roman foot. As they do not refer to the geometers of the middle ages, I cannot guess whence they get their notion, otherwise than from their theory.

I now proceed to demonstrate the existence of this geometrical foot, which I believe to have been the effort of mathematicians to perpetuate and make common what they took to be the Roman foot, on the supposition that it was nothing but the average length of the *human foot*.

Passing over the general expressions of writers who refer to the use of the parts of the body in measurement, and who sometimes distinctly state that the determination of the human foot is necessarily that of the Roman measure, I take first the statement of Clavius, whose term of active life was the latter half of the sixteenth century and who says* very distinctly that the mathematicians, to avoid the diversity of national measures, had laid down a system for themselves. The table of measures which he gives (and dozens of other writers before him) is as follows:

```
    1 breadth of a barleycorn.
   4 =   1 digit.
  16 =    4 =  1 palm (across the middle of the fingers).
  64 =   16 =  4 =  1 foot.
  96 =   24 =  6 =  1½ = 1 cubit.
 160 =   40 = 10 =  2½ = 1¾ = 1 step (gressus).
 320 =   80 = 20 =  5  = 3⅓ = 2 = 1 pace.
 640 = 160 = 40 = 10  = 6¾ = 4 = 2 = 1 perch (pertica).
```

125 paces 1 Italic stadium.
8 stadia 1 Italic mile.
4 Italic miles a German mile.
5 Italic miles a Swiss mile.

The constant reference to this barleycorn measure (which is seldom, if ever, omitted) induced me to try what it would really make. There is some difference between the breadths of barleycorns. A certain statement of Thevenot (cited in the *History of*

---

* ' Enumerandæ sunt mensuræ quibus mathematici, maxime geometræ, utuntur. Mathematici enim, ne confusio oriretur ob diversitatem mensurarum in variis regionibus (quælibet namque regio proprias habet propemodum mensuras), utiliter excogitarunt quasdam mensuras, quæ certæ ac ratæ apud omnes natioues haberentur.' —*Comm. in Sacroboscum.*

Astronomy, Lib. Usef. Kn.) makes the breadths of 144 grains of oriental barley give 1¼ French feet; at which rate 64 would only give 8·53
inches English.   A sample from a London shop gave me (when the
largest grains were picked out) 33 to just more than five inches.
Some other samples, procured from two different parts of England
as the finest which could be got, gave 33 to 5 inches, 33 to 5·1
inches, and 33 to 5·1 inches.   The average of these is 9·8 inches to
64 grain-breadths: a result which coincides more nearly than could
have been expected with the following determinations.

There is a chain of writers who have studied to perpetuate their
geometrical foot by causing a line to be laid down on the page,
representing the digit, the palm, or the foot.   Sometimes the palm
or foot is divided into digits: and of course I rely most on those
whose subdivisions are the best.   As paper is apt to shrink as it
becomes old, the foot deduced from these will be somewhat too small,
and it may be afterwards discussed how much it should be lengthened.   Leaving this for the present, I give the measurements from
different authors.

Margarita Philosophica, above described.   In the Strasburg edition
of fifteen-four, the length of the geometrical palm is less than 2½
inches by from half to three fourths of the 24th of an inch.   Taking
it half way between these, the four-palm foot is 9·9 inches English.
In the Basle edition of fifteen-eight, in which the woodcuts are of
much rougher execution, the palm is 2·64 inches, giving a foot of
10·56 inches.   The palm only is given in both cases.

Oppenheim, fifteen-twenty-four.   Stöffler.   ' Elucidatio Fabricæ
Ususque Astrolabii.'   *Folio in threes* (quarto size).

The digit, palm, and foot, are separately given; the foot is divided
into palms, and all agree excellently well with one another.   The
foot is exactly 9·75 inches English.

Paris, fifteen-twenty-six.   John Fernel.   ' Monalosphærium.'
Folio in threes.

Paris, fifteen-twenty-eight.   John Fernel.   ' Cosmotheoria.'
Folio in threes.

The historical mistake arising out of these works is the most remarkable circumstance attending the loss of the geometrical foot. While
Fernel was publishing the first work, he was meditating (or perhaps
executing) his famous measurement of a degree of the meridian.   In
this first work he lays his geometrical foot down the page, with great
care, as he says (*omni molimine*).   In two copies of this work which I
have examined, the length of the foot is within a sixtieth of an inch
of nine inches and two-thirds, giving 9·65 inches.   In the second work,
in which he announces his measure of the degree, he states that five
of his paces and those of men of ordinary stature make six *geometrical* paces; which, he adds, is agreeable to the opinion of Cam-
panus and others (at least a century and a half before), who made
the mile of 1000 common paces to be 1200 geometrical paces.
Allowing 60 inches (English) to a common pace, which is rather
over than under the truth, this gives a geometrical pace of 50 inches,

and a foot of *ten* inches. This is enough to shew that Fernel was, in the second instance, speaking in rough terms of the foot which he printed *omni molimine* in the first. The geometrical pace being forgotten, and the *monalosphærium* also, the modern historians have assumed that Fernel used the Paris foot: by which he is made to appear to come very near the real degree, whereas he is fifteen miles wrong.

Paris, fifteen-fifty-two. Jac. Koebelius. ' Astrolabii Declaratio.' *Octavo.*

This worthy astrologer, after referring to the perfect notoriety of the system of measures, gives a digit and a palm. The digit is nine-sixteenths of an inch (English), giving a foot of nine inches. The palm is 2½ inches and one-sixteenth, giving 10 inches and a quarter to the foot. The book is small, and the palm incorrectly subdivided.

Frankfort, sixteen-twenty-one. Peter Ryff. ' Questiones Geometricæ.' *Quarto.*

Four very accordant palms are given, indifferently subdivided into digits. Each palm is 2¼ inches and three-sixteenths, giving a foot of 9·75 English inches.

From these different sources, good and bad, we have for the geometrical foot 9·8, 9·9, 10·56, 9·75, 9·65, 10, 9, 10·25, 9·75 inches : the mean is 9·85. But much the best authorities are Fernel and Stöffler, because they are the greatest names, have given the whole foot, and have taken the greatest pains with the subdivisions. Their results are 9·75 and 9·65, with a mean of 9·7.

Taking this as the foot on paper, it remains to ask how much it must be lengthened to allow for the shrinking of the paper. At first, relying on the plate in Dr. Bernard's work on ancient weights and measures, in which the English foot appears to have shrunk by its 42nd part, I was disposed to lengthen the above in the ratio of 41 to 42. But observing that an older English foot, figured in the ' Pathway to Knowledge,' 1596, has shrunk only by its sixtieth part, I am rather inclined to consider the shrinking of Bernard's as an extreme case. And moreover, the two copies of the Monalosphærium give the same foot within one-hundredth of an inch certainly, and less : and it is very unlikely that if the paper had shrunk much, it should have shrunk so equally in two different copies. But, taking one-fiftieth as the outside, it follows that the geometrical foot is any thing the reader pleases between 9·7 and 9·9 English inches. The result from modern barley* gives 9·8, as above shewn.

It is remarkable how completely the English writers are in ignorance of the existence and use of the geometrical foot among their continental neighbours. Blundeville takes Stöffler, in the work above mentioned, to be speaking of a *German* foot, which, says he, Stöffler makes to be 2½ inches less than the English foot.

* Those who have tried to make the *lengths* of three barleycorns into an inch will probably think little of this mode of judging. But I observed that in samples of barley of very different apparent fineness, the difference was in the length of the corns, the breadths hardly varying at all.

Paris, fifteen-fourteen. **Boethius, Jordanus Nemo-rarius, Shirewode** (?), **J. Faber** (Stapulensis). ' In hoc opere contenta Arithmetica . . . Musica . . . epitome . . . Boethii : rithmimachie ludus qui et pugna numerorum appellatur.' *Folio in fours.* Second edition.

The first arithmetic is by Jordanus, the second by Boethius, both edited and commented by Faber. The music (a tract on which was in those days little but a tract on fractions under musical names) is also by Faber. The *Rithmimachia* I suppose to be the work of Shirewode's mentioned in the Introduction, but no author is named. It is a short triple dialogue on the properties of numbers, with some sort of numerical game. Libri attributes it to Faber, probably from its appearance in this edition without an author's name, and headed by an address from Faber, the editor. This second edition was printed by H. Stephens [the first, *Paris,* fourteen-ninety-six, *folio,* was printed by Hopilius, whose partner H. Stephens afterwards was, under the superintendence of David *Lauxius,** of Edinburgh].

Paris, fifteen-fourteen. **Nicolas Cusa.** ' Hæc accurata recognitio trium voluminum operum Clariss. P. Nicolai Cusæ Card. &c.' In three volumes, *folio in fours.*

Cardinal Cusa is put down in several arithmetical lists because one of his *opuscula* is entitled *de Arithmeticis Complementis.* But it is not a work on arithmetic, nor does it even proceed by arithmetic. Or it may be that John Cusa, next mentioned, may have been confounded with Nicolas. For Cusa, see the *Companion to the Almanac* for 1846, p. 14, and *Penny Cyclopædia,* ' Motion of the Earth.'

Vienna, fifteen-fourteen. **Joh. Cusanus.** ' Algorithmus linearis projectilium, de integris perpulchris arithmetrice artis regulis,' &c. *Quarto size.*

A tract of seven pages on counters.

Augsburg, fifteen-fourteen. **Jacob Kobel** (printer). ' Ain Nerv geordnet Rechen biechlin auf den linien mit Rechen pfeningen :' &c. *Quarto, twos and threes* (duodecimo size).

Computation by counters and Roman numerals : the Arabic numerals are explained, but not used. In the frontispiece is a cut representing the mistress settling accounts with her maid-servant by an abacus with counters. This book is said by Kloss to have been also printed by Kobel himself at Oppenheym in the same year.

* *Capiat qui capere potest* is the only general principle on which names like this can be read back into the vernacular. I once found James Hume, well known as a mathematician at Paris in the beginning of the seventeenth century, described as *Scotus Theagrius.* What part of the land of cakes this might be, would probably have eluded all the books in existence : but a Scottish friend, to whom I mentioned my difficulty, solved it at once, by telling me that he must have been one of the Humes of *Godscroft,* a place in which it seems certain Humes are, or were, lords of the soil.

Oppenheym, fifteen-fifteen. **Jacob Kobel.** 'Eyn New geordnet Vysirbüch.' *Quarto in threes.* A work on gauging, in which the Arabic numerals are used.

Vienna, fifteen-fifteen. **Boethius, J. de Muris, Thos. Bradwardin, Nic. Horem, Peurbach, de Gmunden.** 'Contenta in hoc libello. Arithmetica communis, Proportiones breves, de latitudinibus formarum, Algorithmus D. Georgii Peurbachii in integris, Algorithmus Magistri Joh. de Gmunden de minuciis physicis.' *Quarto.*

The name of George Tannstetter as editor of this collection is a fair guarantee for the genuineness of the several works.

The first is an abridgment of Boethius, by John de Muris, of whom elsewhere.

The second is on proportion by Thomas Bradwardin (of the time of Edward III.), who is Bradowardinus, Bragvardinus (as in the work before me), Bragadinus (a confusion perhaps between him and Bragadini), &c. according to the fancy of the speller : Tanner says the book was printed at Paris in 1495. Bradwardine is better known among theologians for his book against Pelagius, than among arithmeticians. He begins thus : 'Proportion is either that which is commonly so called, or properly so called. Proportion commonly so called is the mutual habitude of two things. Proportion properly so called, is the mutual habitude of two things *of the same kind.*' Perhaps the authority of an Archbishop of Canterbury (for he was or was to be nothing less) may induce those who teach the rule of three to remodel their plan according to the archiepiscopal dictum, which is also that of common sense. [Other mathematical works of Bradwardine were printed.]

The third book is on the areas of figures (though nothing to any purpose is done), with the leading idea that the difficulty arises from variation of breadth. Peurbach's work is a summary of operations like that of Sacrobosco ; but though written to explain the Arabic notation, it obviously takes for granted that that notation is generally known. John de Gmunden's work is on sexagesimal fractions (*minuciæ physicæ*).

Paris, fifteen-fifteen. **Gaspar Lax.** 'Arithmetica speculativa magistri Gasparis Lax Aragonensis de sarinyena duodecim libris demonstrata.' Also, same date and place, 'Proportiones magistri Gasparis,' &c. *Folio in threes.*

This is a very diffuse and extensive work, in small black letter. It treats only of the simple properties of numbers ; though it is evident that to a knowledge of Boethius, Lax added that of the arithmetical books of Euclid. The thing which appears most surprising in this and other works of the same kind, is the apparent difficulty of dealing with numbers. Here are upwards of 250 small black-letter pages in folio, filled with propositions on the simplest properties of numbers, and the reader looks in vain for any number so high as 100 used by way of exemplification. For any thing that appears the author could not count as far as 100. Montucla says that Lax was afterwards Pope; but I am assured by a learned Catholic bishop that there never was any Pope of the name.

Lyons, fifteen-fifteen. **John de Lortse** (Dominican).

'Oeuvre tressubtile et profitable de lart et science de arist-meticque : et geometrie translate nouvellement despaignol en francoys Auquel est demonstre par figure evidemment : tant le nombre entier : nombre rompu : regle de compaignies : soubde fin : que toutes aultres choses qui par geometrie et arist-meticque peuvent estre comprises : comme appert par la table cy apres mise.' *Quarto* (O) (166 folios).

This is perhaps the first book in French on commercial arithmetic. I have not found mention of it in any catalogue. It is a good Italian importation, through Spain : and must have made the readers of the Boethian school stare to see what arithmetic really could do. We have thus two Spanish books published at Paris in one year : another (see Siliceus presently mentioned) had been published there the year before.

Venice, fifteen-twenty. **R. Suiseth.** 'Calculator. Sub-tilissimi Ricardi Suiseth Anglici Calculationes noviter emendate atque revise.' *Folio in threes.*

This man is Richard at the beginning of the book, Raymund at the end, while Tanner [and Gesner] call him Roger. His name seems to have been Swinshead, latinised into Swincetus, Suicetus, Suineshevedus, &c. The title of his book, *Calculator,* has given rise to the fiction that he was a great arithmetician, and even al-gebraist ; which, as he lived in the fourteenth century, would have made him a very remarkable man. But nothing can be further from the truth : the book is full of philosophy about intension and remission, and siccity, and calidity and frigidity, and other *quis-quiliæ Suiceticæ,* as some afterwards called them. That number and magnitude are used occasionally, is all that can be said.

Brucker, according to Enfield, Hallam, &c., says this book is as scarce as a white raven. But probably this refers to one of the pre-vious editions. Hain gives three of the fifteenth century, marked 'Padue (no date),' 'Papiæ, 1488,' and 'Papie, 1498.' All these Suissets are *Richards.* (See Tanner *in verb.* 'Swincet;' Wallis *Op.* t. iii. pp. 673, 675, 680, 685.)

Lyons, fifteen-twenty. **Stephen de la Roche Ville-franche.** 'Larismethique nouvellement composee divisee en deux parties dont la premiere tracte des proprietes parfections et regles de la dicte Science,' &c. *Folio in fours* (O).

A large treatise, very full on commercial arithmetic, and containing also geometry. I suppose we should find it largely indebted to the Spanish books above named.

Paris, fifteen-twenty-one. **Boethius.** 'Divi Severini Boetii Arithmetica duobus discreta libris: adjecto commentario, mysticam numerorum applicationem perstringente, declarata.' *Folio in threes.*

The commentary is by Girardus Ruffus, and for absurdity and dulness, it ought to rank high among the mystic commentaries: even to the justification of Cornelius Agrippa, when he asserts that arithmetic is not less superstitious than vain, adding, like a philosopher of his day, that it is only valuable to merchants for the low and mean benefit of keeping their accounts.

London, fifteen-twenty-two. **Tonstall** (Bp. of London). 'De Arte supputandi libri quatuor Cutheberti Tonstalli.' *Quarto.* (Printed by Pynson.)
Paris, fifteen-twenty-nine. The same title. *Quarto in fours.* (Printed by Rob. Stephens.)
Strasburg, fifteen-forty-four. 'De Arte . . . . . Tonstalli, hactenus in Germania nusquam ita impressi.' *Octavo.*

[There is said to have been an earlier Strasburg edition, but I have not seen it: there are several other editions, but no English reprint.] For the life of Tonstall see Anth. Wood, *Ath. Oxon. in verb.* This book was a farewell to the sciences on the author's appointment to the see of London (see the preface): it was published (that is, the colophon is dated) on the 14th of October, and on the 19th the consecration took place. This book is decidedly the most classical which ever was written on the subject in Latin, both in purity of style and goodness of matter. The author had read every thing on the subject, in every language which he knew, as he avers in his dedicatory letter to Thomas More, and had spent much time, he says, *ad ursi exemplum,* in licking what he found into shape. The wonder is, that after this book had been reproduced in other countries, and had become generally known throughout Europe, the trifling speculations of the Boethian school should have excited any further attention. For plain common sense, well expressed, and learning most visible in the habits it had formed, Tonstall's book has been rarely surpassed, and never in the subject of which it treats. It seems to have been very little known to succeeding English writers. P. 419, 426, 427, 439.

Venice, fifteen-twenty-three. **Sacrobosco.** 'Algorismus domini Joh. de Sacro Busco noviter impressum.' *Quarto.*

The edition of this work mentioned by Watt, as by Cirvellus, fourteen-ninety-eight, is a mistake: [it was the work on the sphere which Cirvellus edited in that year] (Hain). Dr. Peacock (p. 416) thinks that this work is attributed to Sacrobosco without sufficient

c

reason, and mentions it only as a manuscript. Mr. Halliwell reprinted it in the *Rara Mathematica*, evidently under the impression that it had never been printed. If it be the work of Sacrobosco, it establishes beyond doubt that he had the Arabic numerals and the method of local value : being nothing but a summary of rules for that arithmetic. The words *noviter impressum* are ambiguous : they may either imply a first impression or any succeeding one, though I have commonly found them used in the latter sense.

I should pause before I rejected as spurious a work which is attributed to Sacrobosco in many manuscript copies extant in different countries, which was printed under his name as early as 1523, and is often cited as his. Dr. Peacock lays stress upon there being no mention of the Arabic system in his other works, those on the sphere and on the calendar. But against the presumption drawn from this circumstance, it may be urged that it was not likely he would introduce mention of a new system of arithmetic in works intended for common use, though he might write a separate work to explain that system. And he would have no motive for alluding to a method which his readers were not acquainted with. There is great probability that Sacrobosco was acquainted with the modern system, which all other evidence goes to shew was introduced into Italy before his time.

It is of course very possible, or, looking at the progress of other things, most probable, that isolated individuals had obtained the Arabic notation from the East, or from Italy, practised it, and written in it, long before it obtained the smallest general currency.

But there is one circumstance which seems to me to lend more than presumption to a still wider supposition, namely, that Arabic notation and rules were known beyond the bounds of Italy, in the time of Sacrobosco, to *more* than a few isolated individuals. Except by importations from the East, it is impossible to say whence the philosophers of the thirteenth century could have got any thing like a set of rules. They certainly had not got any thing Greek, except from the Arabs : the system of Boethius does not give rules of computation. Moreover, the name *algorithm* was in later times so invariably connected with Arabic arithmetic, that the presumption is strong it was so from the beginning. Now Roger Bacon, the contemporary of Sacrobosco, not only used the adjective *algoristicus* several times, but recapitulates the names of a set of rules. The theologian, he says (Jebb, p. 138), should abound in the power of numbering, that he may know all the algoristic modes, not only for integers, but for fractions, to numerate, to add, to subtract, to *mediate* (divide by two), to multiply, to divide, and to extract roots. Now this is precisely the set of rules given in the treatise attributed to Sacrobosco, with one omission : this last treatise distinguishes *duplation* (multiplication by two) from other cases of multiplication, which Bacon does not ; and *progression*, which is, however, only a mixture of other rules.

There is a sentence of Bacon's contemporary, Matthew Paris, quoted by Mr. Hallam, in which he speaks of what can be done with Greek notation, as being what cannot be done either in Latin, or in algorithm, *vel in algorismo.* These words might easily be interpolated; so instead of using them, as Mr. Hallam does, to give ground of presumption that Bacon had the Arabic notation, I should rather use what I have drawn from Bacon to strengthen the genuineness of the words of Paris; and I think Mr. Hallam will be ready to do the same.

In ' Rara Mathematica,' *London,* eighteen-forty-one, *octavo,* second edition, Mr. Halliwell has inserted this treatise of Sacrobosco, together with the poem *de algorismo* attributed to him in some of the manuscripts, but probably enough without foundation. The seven rules above mentioned, are thus given :

> Septem sunt partes, non plures, istius artis;
> Addere, subtrahere, duplareque dimidiare
> Sextaque dividere est, sed quinta est multiplicare
> Radicem extrahere pars septima dicitur esse :

on which Mr. Halliwell quotes a manuscript of perhaps nearly as early a date as Sacrobosco, which describes the seven rules and the necessity of the *letters called figures,* as follows :

En argorismo devon prendre	Et de radix enstracion
Vii especes	A chez vii especes savoir
Adision Subtracion	Doit chascvn en memoire avoir
Doubloison mediacion	Letres qui figures sont dites
Monteploie et division	Et qui excellens sont ecrites

In the same 'Rara Mathematica' are to be found an English manuscript of the 14th century ' On the numeration of Algorism,' and John Norfolk ' In artem progressionis summula,' dated fourteen-forty-five at the end.

Paris, fifteen-twenty-six. **John Martin Siliceus.** ' Arithmetica Joh. Martini Silicei theoricen praxinque luculen-ter complexa' &c. *Folio in threes.*

This is a third edition by Thomas Rhætus; Heilbronner says that the first was *Paris,* fifteen-fourteen, and that the author, who was a professor at Salamanca, named *Gujieno* (Silex), died at Toledo in fifteen-fifty-seven. Montucla says he was Archbishop of Toledo. This is a work of considerable extent of subject, and seems to have no great fault except the usual one of prolixity. It gives a treatise on theoretical arithmetic (Boethian in character), one on the rules of computation, one on the mode of calculating by pebbles or the abacus, and one on fractions. The second edition, fifteen-nineteen, by Orontius Finæus, is in the Royal Society's Library.

Paris, fifteen-twenty-eight. **John Fernel.** ' De pro-portionibus Libri duo.' *Folio in threes.*

A book on proportion of this date is in great part filled with Boethian arithmetic. This book deserves attention : its author had a much better grasp of Euclid than most of his contemporaries.

Leyden, fifteen-thirty-one. **Joach. Fortius Ringel-bergius.** 'De Ratione Studii.' *Octavo* (duodecimo size).

At the end, same place, year, and form, with another title-page, apparently another work, but of this I am not sure, is J. F. R. 'Compendium de scribendis versibus,' to which is attached 'Arithmetica' (O), the most Lilliputian treatise I know of, giving in 17 pages an epitome of Boethian terms, algorithmic rules, and counter calculation. [It is said that the author's name was Sterk.]

Venice, ——. **Giovanni Sfortunati.** 'Nuovo Lume, Libro di Arithmetica . . . . Composto per lo acutissimo prescrutatore delle Archimediane, et Euclidiane dottrine Giovanni Sfortunati da Siena.' *Quarto in fours.*

The end of the work torn out, so I cannot give the date. He mentions L. de Burgo, Calandri, and P. Borgio, and Tartaglia mentions him: which gives certain limits. By the manner in which the author is described, this must be a reprint: it is not likely a work would be originally printed at Venice, one cause of the production of which is stated to be the barbarism of the Venetian dialect. I can find no mention of Sfortunati in catalogues. Cardan, writing in fifteen-thirty-seven, mentions the work of Fortunatus, the same as the above, no doubt. P. 458, 460.

Nuremberg, fifteen-thirty-four. 'Algorithmus demonstratus.' *Quarto.*

John Schoner, the editor, attributes this to Regiomontanus. It is a series of demonstrated propositions of numbers connected with the Arabic notation, involving not only the rules of computation, but such as that the number of figures in the cube cannot exceed three times that in the root, &c. J. Schoner, who edited several writings of Regiomontanus, is likely enough to be right on this one, which is moreover not unworthy of such an author. It is not in the list which he published himself, but it seems to be generally recognised. (Weidler *in verb.*)

Paris, fifteen-thirty-five. **Orontius Fineus.** 'Arithmetica practica libris quatuor absoluta . . . . . Recens ab Authore castigata . . . . . hoc est in nativum splendorem (quem priorum impressorum amiserat incuria) summa fidelitate restituta.' *Folio in fours.*

This must be at least the third edition; Leslie says the first was in fifteen-twenty-five. Orontius died in 1555.

Strasburg, fifteen-thirty-six. **Hudalrich** or **Huldrich Regius.** 'Utriusque Arithmetices Epitome ex variis authoribus concinnata per Hudalrichum Regium.' *Octavo* (duodecimo size).

The first book is on the Boethian arithmetic: the second on the rules of computation with integers and fractions, and on the use of the abacus. There are but 96 small folios with large print : so that if I wished to bring any person into the closest contact with the middle ages at the least expense of reading, with reference both to their mode of expression and operation, I should certainly prescribe this book.

Paris, fifteen-thirty-eight. **Nicomachus** of Gerasa. ' Νι-κομαχου Γερασινου 'Αριθμητικης βιβλια δυο Nicomachi Gerasini Arithmeticæ Libri duo. Nunc primum typis excusi, in lucem eduntur. Parisiis in officina Christiani Wecheli.' *Quarto.*

Nicomachus, who lived about the time of Tiberius, was a Pythagorean, and of course devoted to arithmetic. He was in his day, as Lucian intimates, a Cocker, a person whose name passed as an allusion to reckoning and numbering. The work of Boethius is wholly drawn from him : some people call it a mere translation of, others a comment on, Nicomachus. Perhaps a free rendering might be correct. This edition is Greek without Latin : [there is another, also Greek without Latin, at the end of the *Theologumena Arithmeticæ* attributed to Jamblichus, *Leipsic,* eighteen-seventeen, *octavo* (?)]. It is then ultimately to Nicomachus being a Pythagorean that we owe those once never-ending denominations of numbers and ratios, of which it might now be difficult to trace any vestige in modern language, except the common words *multiple* and *sub-multiple* and the *sesquialter* stop of an organ.

Milan, fifteen-thirty-nine. **Jerome Cardan.** ' Practica Arithmetice et mensurandi singularis.' *Octavo* (O).

This was reprinted, under the title of ' Practica Arithmetica,' in the fourth volume of his collected works, *Leyden,* sixteen-sixty-three, ten volumes, *folio in twos.* On the arithmetic there is no remark to make, except that, as might be expected from an Italian of that day, Cardan shews more power of computation than the French and German writers. There is a chapter recapitulating the numbers which have mystic properties as he calls them, one use of which is in foretelling future events. These are mostly the numbers mentioned in the Old and New Testaments, but not altogether : 400, for instance, makes its appearance, because it was (but it was not) the number of bishops at the Nicene Council. In the tenth volume of the same collection is an unfinished treatise by Cardan, headed ' Artis Arithmeticæ Tractatus de Integris,' which seems to be the commencement of a more extensive treatise than the former one. It may be cited from it that decidedly, in Cardan's opinion, it was Leonard of Pisa who first introduced the Arabic numerals into Europe.

London. 'This boke sheweth the maner of measurynge of all maner of lande, as well of woodlande, as of lande in the felde, and comptynge the true nombre of acres of the same. +

c 2

Newlye invented and compyled by **Syr Rycharde Benese,** Chanon of Marton Abbay besyde London. Prynted in Southwarke in Saynt Thomas hospitall, by me James Nicolson.' *Quarto.*

Again—*London.* 'The Boke of measuryng of Lande as well of Woodland as Plowland, and pasture in the feelde: and to compt the true nombre of Acres of the same. Newly corrected and compiled by **Sir Richarde de Benese.** ¶ Impynted at London, by Thomas Colwell.'

The approximate date of the first we must guess at from the fact that James Nicolson's dated works are from fifteen-thirty-six to thirty-eight, and from the presumption that a printer's undated works come before the dated ones. Thomas Colwell's dated works are from fifteen-fifty-eight to seventy-five. The acre is four roods, each rood is ten *daye-workes*, each daye-worke four perches. So the acre being 40 daye-workes of 4 perches each, and the mark 40 groats of 4 pence each, the aristocracy of money and that of land understood each other easily.* The reprint of the above work wants a set of tables which were in the original, and which I take to be the earliest mathematical tables published in England, except the smaller tables of the same kind which appeared in some acts of parliament. They are of double entry, the entries being made by perches in length and breadth. Thus under 79 opposite 9 we find $\frac{4 \cdot 1}{7 \cdot 3}$ meaning that 79 perches by 9 is 4 acres 1 rood, 7 day-works and 3 square perches. Perhaps this sort of table became common: I find it, without the slightest deviation of form, in Arthur Hopton's ' Baculum Geodæticum sive Viaticum, or the Geodeticall Staffe,' *London,* sixteen-ten, *quarto.*

Paris, fifteen-thirty-nine. **Johannes Noviomagus.** 'De Numeris libri duo.' *Octavo* (small) (O).

Professedly on computation, but interspersed with many curious historical remarks. It contains the first hints I have found of the most probable explanation of the origin of the Roman numerals.

Paris, fifteen-forty-four. **Orontius Finæus.** 'Arithmetica Practica . . . . . multisque accessionibus locupletata.' *Octavo* (O).

---

* More easily than they do at the time when this is written. In the old day, when the aristocracy of money was a Jew, the aristocracy of land used to shut him up, and draw a tooth a day until he protected agriculture to the amount demanded. When Christians became wealthy, a tax upon the use of the teeth seems to have been substituted for their forcible removal. The debates on the abolition of this tax are now proceeding. But, speaking of weights and measures, there is something else which tells a tale about the feasts and Sundays of old England. The sack of wool was 13 tods of 28 pounds each, or 364 pounds. This was arranged, that the tasks of those who spun, &c. might be easily calculated; a pound a day being a tod a month and a sack a year. But where are the Sundays and holidays? Most likely they were made up on the other days.

Paris, fifteen-fifty-five. **Orontius Finæus.** 'De Arithmetica Practica libri quatuor.' *Quarto* (O).

An augmented edition. Had this book been simply headed as arithmetic, I might have let it alone. But as it is called *practical* arithmetic, it must stand as an index of the great advance of the English over their neighbours in computation, as shewn by Tonstall and Recorde. P. 428, 436.

Nuremberg, fifteen-forty-four. **Michael Stifel.** ' Arithmetica integra. Authore Michaele Stifelio. Cum præfatione Philippi Melancthonis.' *Quarto.* (Hutton's Tracts, v. ii. p. 237.)

The two first books are on the properties of numbers, (O), on surds and incommensurables, learnedly treated, and with a full knowledge of what Euclid had done on the subject. The third book is on algebra, and passes for the introduction of algebra into Germany. But Stifel himself, in his preface, acknowledges his obligations to Adam Risen, and professes to have taken all his examples from Christopher Rudolph. Of the latter I can find nothing except a statement that his work was subsequently published: but Kastner (v. i. p. 108) gives some information on the works of Adam and Isaac Riese. P. 424, 436, 474, 450.

Stifel is supposed to have been the inventor of the signs + and − to denote addition and subtraction : and the manner in which he speaks of these signs seems to imply that they were of his own introduction (p. 110). It is " *we place* this sign," &c. and " *we say* that the addition is thus completed;" and so on. The sign +, in the hands of Stifel's printer, has the vertical bar much shorter than the other : and when it is introduced in the woodcuts by the engraver, the disproportion is greater still. One would (supposing the wood-engraver to have imitated the handwriting) have supposed that − was first used, and that + was derived from it by putting a small cross-bar for a distinction. This form is rather against the late learned Professor Rigaud's idea (see Mr. Davies' *Solutions of Hutton's Mathematics,* p. 11) that + is a corruption of P, the initial of *plus,* and also against Mr. Davies' ingenious conjecture that it is a corruption of *et* or *&.* Stifel does not call the signs *plus* and *minus,* but *signum additorum* and *signum subtractorum.* In no instance do I find him using the first pair of appellatives at all.

I had written the above, and become fully impressed, by repeated examination, with the suspicion that the vertical bar in + was, in Stifel's invention, a small distinction superadded on the sign −, when I happened, in the immediate preparation of these sheets for press, to remember Colebrooke's statement that the Hindoos (our original teachers in algebra) use a dot for subtraction, and nothing but the absence of the dot for addition. It may be likely then, that in the first instance the Hindoo dot was elongated into a bar, to signify subtraction, addition having no sign: and that the first who found it convenient to introduce a sign for addition, merely adopted the sign for subtraction with a difference.

Against the above conjecture it may be argued, that the European algebra comes from India through the Mahometan writers, who use no signs at all. On this I can only remark at present, that I have long suspected, from many circumstances, that there was a more direct communication with India in the introduction of algebra than any one now believes.

Dr. Ritchie suggested that perhaps + was used to denote addition as being two marks joined together, and made significative of two numbers joined together: and that − denoted subtraction, as indicating what is left after one of the marks of the junction is removed. This suggestion is more like what *ought to have been* the derivation than any I have seen: but there is little reason to suppose that any such attempt at significant symbols would have been made in the sixteenth century.

M. Libri attributes the invention of + and − to Leonardo Da Vinci, in whose manuscripts he says he had seen them. But as it happened to me to discover, in a manuscript of that justly celebrated man (if it be not more correct to say *unjustly* celebrated, as not celebrated enough for his merits) now in the British Museum, the sign + used for the figure 4, I think the question must rest open until citations which establish the point without mistake can be produced.

No place, nor date. **Wm. Buckley.** 'Arithmetica memorativa sive compendaria Arithmeticæ tractatio.' Apparently *octavo*.

This book has been described both by Leslie and Peacock, who have severally quoted some of the verses of which it consists. They say it was first printed in fifteen-fifty, and afterwards at the end of Seton's Logic in sixteen-thirty-one. I should probably not have seen it to this day (for after Leslie's full citations, I should hardly have made active search for it), had not an anonymous correspondent (whom I have only this way of thanking) obligingly forwarded to me an imperfect copy which he happened to find among his papers. It is, I suppose, the extract from Seton's Logic: the signature in the title-page is Q 2, after a *libellus ad lectorem*, 'Quisquis Arithmeticam . . . . qui meminisse dedit:' and there is a preface signed T. H. It is worth noting that Buckley's Latin name, *Buclæus*, has been frequently written *Budæus* by the continental bibliographers, so that in all probability the learned author of the book *De Asse* has been taken for the author of these verses. Hylles calls him *Buckle*. P. 437.

[According to Watt, there were editions of Seton's Logic, with 'Huic accessit ob artium ingenuarum inter se cognationem Arithmetica Gulielmi Buclæi,' *London*, fifteen-seventy-two, seventy-four, and seventy-seven, *octavo*.]

Basle, fifteen-fifty. **John Scheubel.** 'Euclidis Megarensis . . . . . Algebræ porro Regulæ . . . . . his libris præmissæ sunt . . .' *Folio in twos*.

The algebra was reprinted, *Paris*, fifteen-fifty-two. *Quarto*. 'Algebræ compendiosa facilisque descriptio,' &c. For a description of the work see Hutton's Tracts, vol. ii. p. 241.

Leyden (*Lugduni*, without further description), fifteen-fifty. **Henry Cornelius Agrippa.** ' De Occulta Philosophia Libri Tres.' *Octavo* (Italic letter).

As the preface is dated *Mechlin*, fifteen-thirty-one, and [an edition appeared there in fifteen-thirty-three], this latter must be the first. This book has as much right in my list as that of Jordano Bruno presently noticed, treating of numbers in much the same way. It is a property of seven that there are seven angels who stand before God; of twelve, that there are twelve apostles, twelve permutations of the letters of the tetragrammaton, &c. A good many of the more formal works on arithmetic of this period abound in enumerations of the same kind, evidently considered as appertaining to the higher uses of the science.

Augsburgh, fifteen-fifty-four. **Psellus.** ' De quatuor scientiis mathematicis.' Edited by Xylander. Gr. Lat. *Octavo* (duodecimo size).

Psellus lived in the ninth century. The part relating to Arithmetic is a mere compendium of arithmetical terms, perhaps from Boethius.

Venice (the first two parts fifteen-fifty-six, the rest fifteen-sixty). **Nicolas Tartaglia.** ' La prima parte dei numeri e misure aritmetica geometria et algebra.' *Folio in threes* (O).

Of this enormous book I may say, as of that of Pacioli, that it wants a volume to describe it. P. 430, 450. It was partly translated into French as follows: *Paris*, sixteen-thirteen. ' L'Arithmetique de Nicolas Tartaglia, grand mathematicien et prince de praticiens, recueillie et traduite d'Italien en Francois par Guillaume Gosselin de Caen.' *Octavo.*

London, fifteen-fifty-seven. **Robert Recorde.** ' The whetstone of witte, whiche is the seconde parte of Arithmetike : containyng thextraction of Rootes : The Cossike practise, with the rule of Equation : and the woorkes of Surde Numbers.' *Quarto.*

On this see the *Companion to the Almanac* for 1837, p. 36; Hutton's Tracts, v. ii. p. 243; P. 369. It is rarely remembered that the old name of Algebra, the *cossic* art (from *cosa,* thing), gave this first English work on algebra its punning title the *whetstone* of wit, *Cos ingenii.*

Lione, fifteen-fifty-eight. **Giov. Franc. Peverone.** ' Due

brevi e facili trattati. Il primo d'Arithmetica : l'altro di Geo-metria,' &c. *Quarto.*
A book of the very simplest examples of computation.

Leyden, fifteen-fifty-nine. **Joh. Buteo.** 'Joan. Buteonis Logistica quæ et Arithmetica vulgo dicitur.' *Quarto in fours.* (O).

In the same year was published, by the same author, his *Opera Geometrica,* the first of which is a plan of Noah's ark, with calcula-tions as to its sufficiency for holding all the animals and their pro-vender. P. 437.

Paris, fifteen-sixty. **James Pelletier.** ' Jacobi Pele-tarii Cenomani, de occulta parte numerorum, quam algebram vocant, libri duo.' *Quarto.* [First edition, fifty-eight.]

This is said to be the first French work on algebra. There is little purely arithmetical in it: at the end is a table of squares and cubes up to those of 140. See Hutton's Tracts, vol. ii. p. 245.

London, fifteen-sixty-one. **Robert Recorde.** ' The Grounde of artes : Teaching the worke and practise of Arith-metike, both in whole numbres and Fractions, after a more easyer and exacter sorte than any like hath hitherto been sette forthe : Made by M. Robert Recorde, Doctor of Physik, and now of late ouerseen and augmented with new and necessarie Additions.

I[ohn] D[ee].

All youth and Elde that reasons Lore
Within your breastes will plant to trade
Of Numbers might the endles store
Fyrst vnderstand, than farther wade'

Octavo (duodecimo size) (O).

See the *Companion to the Almanac* for 1837, p. 32. That this book was published about fifteen-forty there is internal evidence: and Tanner gives it that date. Dr. Peacock has fifteen-forty-two. But I have never seen any edition earlier than this. John Dee published it again, *London,* fifteen-seventy-three, *octavo,* with other verses in the title-page, and the original dedication to Edward VI. restored. The preceding edition, though published under Elizabeth, was pro-bably arranged in the life of her sister, which may account for the omission of the dedication. There have been many editions. Hart-well's edition of Mellis's edition was published thrice at least, in sixteen-forty-eight and sixteen-fifty-four, and once more with the date torn out in our copy. Mellis added a third part, on practice and other things: in Roman letter, the sacred text of Recorde hav-ing always its old black letter. The last edition I know of it is by Edw. Hatton, *London,* sixteen-ninety-nine, *quarto;* it has an additional book called 'Decimals made easie.' P. 408-10, 434, 454, 476.

Vienna, fifteen-sixty-one. **Christoffer Rudolff.** ' Kunstliche rechnung mit der ziffer und mit der zalpfenningen,' &c. *Octavo* (small).

This is the man to whom Stifel refers as his instructor in algebra. But there is nothing like either of the signs + or − in his book : so that Stifel does not appear to have got them from him. This book is not algebra : and whether any thing of Rudolph's deserving of that name was published is more than I can now settle.

Antwerp, fifteen-sixty-five. **Valentine Menher de Kempten.** ' Practicque pour brievement apprendre à Ciffrer, et tenir livre de Comptes, avec la regle de Coss, et Geometrie.' *Octavo.* [Date at the end, sixty-four, July.]

Here are books on Arithmetic, Book-keeping, Algebra, Geometry, and Trigonometry : it appears that part of the arithmetic had been published in fifteen-fifty-six, of the geometry in fifteen-sixty-three, of the algebra in fifteen-fifty-six. The arithmetic is an early specimen of the book of mere rules. In this and other French books of the same date, and even much later, the words *septante* and *nonante* are used for seventy and ninety. The book-keeping is a short treatise, or rather example, of double entry. The algebra is more extensive, on the model of Stifelius, with many questions leading to simple equations, each embellished with an illustrative picture in wood. The geometry is mensuration. The spherical trigonometry is from Regiomontanus, and refers to no other tables. This book has some pretension to be called a course of mathematics. Menher is often quoted in England for his tables of rebate or discount.

Antwerp, fifteen-sixty-seven. **Pedro Nunez.** ' Libro de algebra en arithmetica y geometria.' *Octavo* (small).

This book is wholly algebraical. See Hutton's Tracts, vol. ii. p. 250.

London, fifteen-sixty-eight. **Humfrey Baker.** ' The Well Spryng of Sciences, which teacheth the perfecte woorke and practise of Arithmeticke . . . . beautified with moste necessary rules and questions . . . .' *Octavo* (very small size) (O).

The reprints will be presently mentioned.

Paris, fifteen-seventy-one. **Alex. Vandenbussche.** ' Arithmetique militaire.' *Quarto.*

A short work, mostly on the arithmetic of the arrangement of troops, with a large number of military maxims and observations.

The first book ends with a question described as *non moins plaisante que belle*, as follows. A *soldat désespéré* asked of an *arithméticien fantastique* how far it was from Milan to hell. The mathematician, placing it at the centre of the earth, made the distance 1145¼ leagues. This, and *Je prie à Dieu nous garder d'y aller*, is the whole of the joke.

London, fifteen-seventy-one. **Leonard Digges.** 'A Geometricall Practise, named Pantometria . . . . framed by Leonard Digges, Gentleman, lately finished by Thomas Digges his sonne. Who hath also thereunto adjoyned a Mathematicall Treatise of the five regulare Platonicall bodies,' &c. *Quarto*. Reprinted as follows : *London*, fifteen-ninety-one. 'A Geometrical practical treatize named Pantometria. . . . . First published by Thomas Digges Esquire . . . . lately reviewed by the author himself,' &c. *Quarto*.

Almost the whole is application of Arithmetic to Geometry, in measurement both of planes and solids.

In this book I find the earliest printed mention I ever met with of the *theodelite :* a word the derivation of which has long been a puzzle. It is known to be an exclusively English word in all time preceding the middle of the last century. Digges calls it 'the circle called *theodelitus*.' A moveable radius travelling round a circle, for the measurement of angles, had been long denoted by the Arabic word *alhidada* (whence the French word *alidade*). *Theodelite* seems unlikely to be a corruption of this, at first : nor should I have suspected such a thing, if I had not found in Bourne's 'Treasure for travailers,' *London*, fifteen-seventy-eight, *quarto*, the intermediate formation, *athelida*, used for the same thing. I hold it then pretty certain that this is the true origin of the word *theodelite*, which was so spelt, not *theodolite*. (See the *Philosophical Magazine*, April, 1846.)

Venice, fifteen-seventy-five. **Franc. Maurolycus.** 'Arithmeticorum libri duo.' *Quarto in fours*.

On the properties of numbers and the doctrine of incommensurables ; a superior work to the mass of those which then treated of similar subjects.

Basle, fifteen-seventy-nine. **Christian Urstisius.** 'Elementa Arithmeticæ, logicis legibus deducta.' *Octavo* (duodecimo size) (O).

It is said that Urstisius was a follower of Copernicus and a teacher of Galileo. This Arithmetic was translated by T. Hood, *London*, fifteen-ninety-six, *octavo* (small). 'The Elements of Arithmeticke most methodically delivered,' &c.

Brescia, fifteen-eighty-one. **Ant. Maria Vicecomes** (Visconti). 'Practica numerorum et mensurarum,' &c. *Quarto*.

A book of mixed arithmetic and algebra, in which the modern method of division is used. It would, I think, repay an examination, particularly on the part which relates to the extraction of roots.

Antwerp, fifteen-eighty-one. **Gemma Frisius.** ' Arithmeticæ practicæ methodus facilis, in eandem J. Steinii et J. Peletarii annotationes,' &c. *Octavo* (small size) (O).

This is a reprint, much augmented, of a compendious work of Frisius, the ' Ar. pract. meth. fac.' *Witteberg,* fifteen-sixty-three, *octavo* (small). But there is an earlier edition of Gemma Frisius, also by Peletarius, being the earliest I have met with, as follows :

Paris, fifteen-sixty-one. ' Arithmeticæ Practicæ Methodus facilis, per Gemmam Frisium . . . . jam recens ab ipso authore emendata, et multis in locis insigniter aucta. Huc accesserunt Jacobi Peletarii Cenomani annotationes . . . . Quibus demum ab eodem Peletario additæ sunt Radicis utriusque demonstrationes.' *Octavo* (small) (O).

These demonstrations of the rules for the square and cube roots Peletarius treats as a great boon to his reader : and states that he was very near contenting himself with a reference to his book on *occult* properties of numbers (his Algebra of 1560, before mentioned).

London, fifteen-eighty-three. **H. Baker.** ' The Well spring of sciences. Which teacheth the perfect worke and practise of Arithmeticke . . . . set forthe by Humfrey Baker, Londoner, 1562. And nowe once agayne perused augmented and amended in all the three partes, by the sayde Aucthour.' *Octavo* (duodecimo size).

Reprinted (same place and size) in sixteen-fifty-five, with a reference at the end to an edition of sixteen-forty-six. It is one of the books which break the fall from the ' grounde of Artes' to the commercial arithmetics of the next century. There are some short rules for particular cases, and great attention to the rule of practice. Among the peculiarities of the book is a notion, apparently, that none but fractions should deal with fractions : for Baker will not double $\frac{3}{5}$, for instance, by multiplying by 2, but only by dividing by $\frac{1}{2}$. P. 452.

London, fifteen-eighty-four. **John Blagrave.** ' The mathematical jewel.' *Folio in twos.*

See the *Companion to the Almanac* for 1837, p. 41. This book is curious, because the woodcuts are all done by the author's own hand, and Walter Venge, the printer, is not known to have printed any other work. [For Blagrave's legacy to the three parishes of Reading, see Ashmole's Berkshire, v. iii. p. 372 : for the appearance of his ghost, "credibly reported by many honest and discreet persons," see the *Annus Mirabilis*, 1662, p. 49.] This book is not arithmetical, but the contrary : it is an avowed attempt to drive computation out of astronomy by the introduction of an instrument called the *Jewel*, which is a projection of the sphere. Of the tables of sines he says, ' But now there are in like maner an infinite number of intricate questions more harder far than any yet propounded, which by Regiomont. Copernicus and others doctrine, grow to great toile with their Synes, calculations, and proportions, wherein first they hunt about for one Syne, which they call *inventum primum*, then for another, which they call *inventum secundum*, and commonly *inventum tertium*, and perhaps *quartum*, and so foorth. After all these are found, then they multiplie and devide them, and compare their proportions : and when al is done, that they have founde the Seyne sought for : they yet are faine to goe to their tables for the arch correspondent.' There was a great disposition at this time among the English to try to substitute instruments for computation. Blagrave's Jewel stood in high estimation ; and as late as 1658, John Palmer, who published a description of it under the name of the *Catholic Planisphere*, says it was a subject of frequent inquiry where the instrument and the description were to be met with.

Leyden, fifteen-eighty-five. **Simon Stevinus.** 'L'Arithmetique de Simon Stevin de Bruges : Contenant les computations des nombres arithmetiques ou vulgaires : Aussi l'algebre, avec les equations de cinc quantitez. Ensemble les quatre premiers livres d'Algebre de Diophante d'Alexandrie maintenant premierement traduicts in Francois. Encore un livre particulier de la Pratique d'Arithmetique, contenant entre autres, Les Tables d'Interest, La Disme ; Et un traicté des Incommensurables grandeurs : Avec l'Explication du Dixiesme livre d'Euclide.' *Octavo*.

Leyden, sixteen-thirty-four. **Albert Girard.** Les Oeuvres Mathematiques de Simon Stevin de Bruges,' &c. *Folio in threes*.

See Hutton, Tracts, vol. ii. p. 257, and Peacock, Enc. Metr. Arithmetic, p. 440. My copy of the first work is imperfect, stopping at the end of the Diophantus; but as Alb. Girard gives every thing in the collected works, just in the order announced in the above title-page, I suppose there can be no doubt that every thing there announced was published in 1585. The *Disme* is the first announcement of the use of decimal fractions. Dr. Peacock supposes it was first published in Dutch about 1590 : we here see that it was published in French in 1585. Hutton speaks of a Dutch edition in 1605. I do not know where either date* was got from. On looking at the verses with which Stevinus's friends (according to cus-

* Perhaps 1605 is a mistake for 1626, at which date [R. Soc. Libr. Cat.] De Thiende, leerende alle rekeninghen, &c. was published at Gouda.

tom) have eulogised him, a part of a book to which I am now and
then under obligation, I find that the *Disme*, &c., was attached to
the edition of 1585, with a new paging. Jagius Tornus, *philomathes*,
be he whom he may, quotes and writes as follows, with a side-note:

> Non fumum ex fulgore sed ex fumo dare lucem
> Cogitat, ut speciosa dehinc miracula promat.
> Sume unum è multis. quid non Decarithmia præstat   Decarithmia
> Divinum scriptoris opus? cui non ego si vel   La Disme
> Aurea mi vox sit, centum linguæ, oráque centum   pag. 132.
> Omni ætate queam laudes persolvere dignas.

Accordingly, assuming Girard to have faithfully copied his original
in his headings (which he has done as far as my means of comparison go), the method of decimal fractions was announced before
1585, in Dutch.

The characteristic of Stevinus is originality, accompanied by
a great want of the respect for authority which prevailed in his
time. For example, great names had made the point in geometry
to correspond with the unit in arithmetic: Stevinus tells them that 0,
and not 1, is the representative of the point. And those who cannot
see this, he adds, may the Author of nature have pity upon their
unfortunate eyes; for the fault is not in the thing, but in the sight
which we are not able to give them. P. 426, 440, 460.

[Dr. Peacock (p. 440) mentions an English translation of this
tract by Richard Norton, sixteen-eight, under the title 'Disme, the
arte of tenths, or decimal Arithmetike . . . . . invented by the excellent mathematician, Simon Stevin.']

Since writing the above I have examined the account of Stevinus
given by M. Quetelet, presently referred to. It is stated that the
first work of Stevinus was his table of interest, *Antwerp*, fifteen-
eighty-four ——; this was reprinted the next year in the Arithmetic.
In the Supplement of the *Penny Cyclopædia*, 'Tables,' I have noted,
from the contents of this table and of the *Disme*, the great probability that the table of compound interest *suggested decimal fractions:*
the only doubt was, which was written first, the table or the Disme.
The statement of M. Quetelet clears this up: and I hold it now next
to certain that the same convenience which has always dictated the
decimal form for tables of compound interest was the origin of decimal fractions themselves.

London, fifteen-eighty-eight.   **John Mellis.** 'A briefe
instruction and maner hovv to keepe bookes of Accompts after
the order of Debitor and Creditor, and as well for proper
Accompts partible, &c. By the three bookes named the Me-
moriall Journall and Leager, and of other necessaries appertaining to a good and diligent marchant. The which of all
other reckoninges is most lawdable: for this treatise well and
sufficiently knowen, all other wayes and maners may be the
easier and sooner discerned, learned and knowen. Newely
augmented and set forth by John Mellis Scholemaister, 1588.

Imprinted at London by John Windet, dwelling at the signe of the white Beare, nigh Baynards Castle. 1588.' *Octavo.*

This is the earliest English book on book-keeping by double entry which has ever been produced. At the end of the book-keeping is a short treatise on arithmetic (O). But Mellis says : ' Truely I am but the renuer and reviver of an auncient old copie printed here in London the 14 of August, 1543. Then collected, published, made and set forth by one Hugh Oldcastle Scholemaster, who, as appeareth by his treatise then taught Arithmetike and this booke, in Saint Ollaves parish in Marke Lane.'

[Watt and others (but not Ames) mention a *quarto* treatise having this date, fifteen-forty-three, and entitled ' A profitable Treatyce called the Instrument or Boke to learne to knowe the good order of the kepyng of the famouse reconynge, called in Latyn, Dare and Habere, and in Englyshe, Debitor and Creditor.' But no author's name is given.]

Beckman quotes Anderson as to a book of James Peele, *London,* fifteen-sixty-nine, *folio,* on book-keeping: but it may be doubted whether this was not a book on single entry.

London, fifteen-ninety. **Cyprian Lucar.** ' A treatise named Lucarsolace,' &c. *Quarto.*

A work on mensuration, interspersed with arithmetical rules. It contains the description of a fire-engine. We find from it that the surveyors used, not black-lead, but ' fine-pointed coles or keelers,' of which three or four might be bought of any painter for a penny. Lucar also translated part of Tartaglia on Artillery, with a long appendix. *London,* fifteen-eighty-eight. *Folio in threes.*

London, fifteen-ninety. **Thomas Digges.** ' An Arithmetical warlike treatise named Stratioticos .... first published by Thomas Digges Esquire, anno Salutis, 1579,'&c. *Quarto* (O).

There is here a brief and good treatise on arithmetic, and some algebra of the school of Recorde and Scheubel : but the greater part of the work is on military matters.

Helmstadt, fifteen-ninety. **Heizo Buscher.** ' Arithmeticæ libri duo.' *Octavo* (small) (O).

A short and very backward work.

Bergomi, fifteen-ninety-one. **Peter Bungus.** 'Numerorum Mysteria.' Second edition. *Quarto in fours.*

Dr. Peacock gives some description of this fantastical work, page 424.

Frankfort, fifteen-ninety-one. **Jordano Bruno.** 'De Monade numero et Figura liber Consequens Quinque de Minimo magno et Mensura. Item de Innumerabilibus, immenso, et Infigurabili; seu De Universo et Mundis libri octo.' *Octavo* (duodecimo size).

On Jordano Bruno see Bayle, Drinkwater-Bethune's Life of Galileo, p. 8, and Libri, vol. iv. p. 141, &c. who makes him a vorticist before Des Cartes, an optimist before Leibnitz, a Copernican before Galileo. The end of it was that he was roasted alive at Rome, February 17, 1600, at the age of fifty: ostensibly for heresy (he had been a Dominican), but most probably in revenge for his satirical writings. He does not seem, however, to have been a Protestant martyr; as his opinions would probably have procured his death in almost any state in Europe, as matters were at that time. Bruno, says M. Libri, seems to have become a Copernican by a sort of intuition, for he was any thing rather than a mathematician. His book on monad and number fully confirms this last statement: it is a collection of dissertations on individual numbers. But it has the advantage over Pacioli and Bungus (P. 424), that the Latin is better, and a great part of it is in verse. In the *triads*, for instance:

> Efficiunt totum Casus, Natura, Voluntas,
> Dat triplicem mundum Deitas, Natura, Mathesis,
> Hinc tria principia emanant, Lux, Spiritus, Unda,
> Est animus triplex Vitâ, Sensu, Ratione.

M. Libri has quoted at length (vol. iv. note x.) the Copernican chapter of the work *de Immenso*.

Frankfort, fifteen-ninety-two. **Peter Ramus and Lazarus Schoner.** 'Petri Rami Arithmetices Libri Duo, et Algebræ totidem: a Lazaro Schonero emendati et explicati. Ejusdem Schoneri libri duo: alter, De Numeris figuratis; alter, De Logistica Sexagenaria.' *Octavo* (O).

The first edition of Ramus is said to be of fifteen-eighty-four. P. 427.

London, fifteen-ninety-two. **Thomas Masterson,** 'his first booke of Arithmeticke.'
London, fifteen-ninety-two. **Thomas Masterson,** 'his second booke of Arithmeticke.'
London, fifteen-ninety-four. **Thomas Masterson,** 'his Addition to his first booke of Arithmetick.'
London, fifteen-ninety-five. **Thomas Masterson,** 'his thirde booke of Arithmeticke.' All *Quarto* (O).

The first book and the addition are on abstract numbers and fractions: the second book is on commercial arithmetic: the third book is on the extraction of roots, on surds, and on cossic numbers, or algebra. Masterson must have been a valuable help to the student: though he would have been more so if he had used the modern method of division.

London, fifteen-ninety-four. **Blundevile.** ' M. Blundevile His Exercises, containing sixe Treatises,' &c. *Quarto in
fours.*

He was the first introducer of a complete trigonometrical canon
into English, the first announcer of Wright's discovery of meri-
dional parts, and he exercised much influence on the studies of
the first part of the seventeenth century. Nor is he altogether
out of date yet; for I lately observed (*Mechanic's Magazine,* vol.
xliv. p. 477) that a patent for an improvement in horse-shoes was
upset in the Court of Chancery, on proof that Blundevile had described it in one of his books on horses, to which he refers
several times in the present work. The first of these exercises is a
treatise on arithmetic, in the form of question and answer, written
for Elizabeth Bacon, the sister of the great philosopher. It is a
dogmatical treatise, sufficiently clear. The seventh edition of these
exercises was published in sixteen-thirty-six, and I believe there
was none later.

London, fifteen-ninety-six. ' The Pathway to Knowledge.
Conteyning certaine briefe Tables of English waights, and Measures . . . . With the Rules of Cossicke, Surd, Binomicall, and
Residuall Numbers, and the Rule of Equation, or of Algebere
. . . . And lastly the order of keeping of a Marchants booke,
after the Italian manner, by Debitor and Creditor. . . . Written
in Dutch, and translated into English, by W. P.' *Quarto* (O).

This work has escaped the notice of Ames, and was perhaps
confounded with the geometrical work of the same name by Recorde.
The English translator's preface, giving an account of the existing
weights and measures, is the most complete thing there is of the
day. The arithmetic is not equal to that of Recorde or Masterson,
nor the book-keeping to that of Mellis, and the algebra is far behind Recorde's. The translator gives the following verses, the first
of which are now well known. I suspect he is the author of them,
having never seen them at an earlier date : Mr. Halliwell, who is
more likely than myself to have found them if they existed very
early, names no version of them earlier than 1635 :

> Thirtie daies hath September, Aprill, June, and November.
> Febuarie eight and twentie alone, all the rest thirtie and one.

> Looke how many pence each day thou shalt gaine,
> Just so many pounds, halfe pounds and groates :
> with as many pence in a yeare certaine,
> Thou gettest and takest, as each wise man notes.

> Looke how many farthings in the weeke doe amount.
> In the yeare like shillings, and pence thou shalt count.

To complete this subject : Mr. Davies [Key to Hutton's Course,

p. 17] quotes the following from a manuscript of the date 1570, or near it:

> Multiplication is mie vexation,
> And Division is quite as bad,
> The Golden Rule is mie stumbling stule,
> And Practice drives me mad.

Alcmaar (preface dated from Amsterdam), fifteen-ninety-six. **Nicolaus Petri.** 'Practicque omte Leeren Rekenen Cypheren ende Boeckhouwen met die Reghel Coss . . . .' *Octavo* (small) (O).

Contains arithmetic, algebra, geometry, and an example of book-keeping.

Venice, fifteen-ninety-nine. **Joh. Bapt. Benedictus.** 'Speculationum Liber.' *Folio in twos.*

The first speculation is entitled 'Theoremata Arithmetica,' and is a laboured explanation of the principles of arithmetical rules by reference to geometry, beginning with the old difficulty of the product of two fractions being less than either of the factors.

London, sixteen-hundred. **Thomas Hylles.** 'The arte of vulgar arithmeticke, both in integers and fractions, devided into two bookes . . . . Nomodidactus Numerorum . . . . Portus Proportionum . . . . whereunto is added a third book entituled Musa Mercatorum, comprehending all rules used in the most necessarie and profitable trade of merchandise . . . . Newly collected, digested, and in some part devised, by a wel willer to the Mathematicals.' *Quarto* (O).

This is in dialogue; but whenever any rule or theorem is delivered, it is in verse. It is a big book, heavy with mercantile lore. The following are specimens of the verses :

> Number is first divided as you see
> For number abstract, and number contract
> And numbers abstract are such as stand free
> From every substance so cleare and exact
> That they have no sirname or demonstration
> Save the pure units of their numeration.

> All primes together have no common measure
> Exceeding an ace which is all their treasure.

> Addition of fractions and likewise subtraction
> Requireth that first they all have like basses
> Which by reduction is brought to perfection
> And being once done as ought in like cases,
> Then adde or subtract their tops and no more
> Subscribing the basse made common before.

The partition of a shilling into his aliquot parts.

> A farthing first findes fortie eight
> An halfepeny hopes for twentie foure

> Three farthings seekes out 16 streight
> A peny puls a dozen lower.
> Dicke dandiprat drewe 8 out deade
> Twopence tooke 6 and went his way
> Tom trip and goe with 4 is fled
> But goodman grote on 3 doth stay
> A testerne only 2 doth take
> Moe parts a shilling cannot make.

In spite of all this trifling, Hylles was a man of learning. He cites both *Lucas* de Burgo and *Peter* de Burgo.

Paris, sixteen-hundred. **John Chambers.** 'Barlaami Monachi Logistica nunc primum Latiné reddita et scholiis illustrata.' *Quarto*.

The Greek is given at the end. Barlaam lived in the fourteenth century, and his work is mostly on fractions and on proportions. See the *Penny Cyclopædia*, 'Barlaam,' for information on the author and on this edition.

Leyden, sixteen-three. **C. Dibuadius.** 'In Geometriam Euclidis prioribus sex Elementorum libris comprehensam Demonstratio Numeralis.' *Quarto*.

The six books of Euclid are, for the most part, here verified by arithmetical or trigonometrical instances. The usual mode, or *linear* demonstration, is made to follow in another book.

Leyden, sixteen-five. **Simon Stevinus.** 'Tomus secundus Mathematicorum hypomnematum de Geometriæ praxi.' *Folio in threes*. (The printing delayed beyond the title-date.)

This work contains the celebrated Latin treatise on book-keeping (Peacock, *Encyl. Metrop.* 'Arithmetic,' p. 464), in which the difficulty of finding Latin renderings of the technical terms is well got over. All writers attribute this translation to the celebrated Willebrord Snell, who passed from fourteen to eighteen years of age while the work was printing. Now the fact is, that not only does Stevinus date his own preface to this treatise on book-keeping in August 1607, but at the end of the volume he excuses himself from giving several matters which had been announced at the beginning, because the printer was tired of waiting, and he had not satisfied himself about them. And in this same final notice he mentions *Will. Sn.* (Snell) as a person who had written a letter for him : styling him *mathematum neque ignarus neque expers;* much too poor a compliment for a youth who, at such an age, had translated, and therefore understood, all the writings of Stevinus. The story about the translation is from Gerard Vossius, who was afterwards Snell's colleague, and who made a mistake and confusion of persons which was copied by Bayle, Beckmann, &c. M. Quetelet has recently published a notice of Stevinus, which states that he was born in 1548, and died in 1620. P. 464.

Augsburgh, sixteen-nine. **Geo. Henischius.** 'Arithmetica perfecta et demonstrata.' *Quarto* (ON).

A laboured work in seven books. Its algebra is that of a former day; its power of computation very fair. Dr. Peacock refers several times to another work of Henischius, ' De numeratione multiplici,' sixteen-five.

Mayence, sixteen-eleven. **Christopher Clavius.** 'Epitome Arithmetices Practicæ.' *Folio in threes* (O).

This is in the second volume of Clavius's collected works. [It is said to have been first published in fifteen-eighty-three.] Perhaps there are more extensive examples of the square root worked by the old method in this treatise than would easily be found elsewhere. P. 426.

Bologna, sixteen-thirteen. **Pietro Antonio Cataldi.** ' Trattato del modo brevissimo di trovare la Radice quadra delli numeri.' *Folio.*

The rule for the square root is exhibited in the modern form, and Cataldi shews himself a most intrepid calculator. But the greatest novelty of the book is the introduction of continued fractions, then, it seems, for the first time presented to the world. Here, with great labour, but still with success, Cataldi reduces the square roots of even numbers to continued fractions of the form $a + \dfrac{m}{n+} \dfrac{m}{n+}$ &c. He then uses these fractions in approximation, but without the assistance of the modern rule by which each approximation is educed from the preceding two. Thus he reduces, among other examples, the square root of 78 to

$$8\frac{307376382593588549068368 1}{369548983412298550212624 0}$$

London, sixteen-thirteen. **John Tap.** ' The Pathway to Knowledge,' &c. *Octavo* (O).

This professes to be a reprint of the work published under that name (1596), and above described. But it is in fact the substance of the above work thrown into the form of dialogue, with the book-keeping reprinted, and some algebra taken from V. Menher above mentioned.

London, sixteen-thirteen. **Richard Witt.** 'Arithmeticall questions, touching the Buying or Exchange of Annuities ; Taking of Leases for Fines, or yearly Rent ; Purchase of Fee-Simples ; Dealing for present or future Possessions ; and other Bargaines and Accounts, wherein allowance for disbursing or forbeareance of money is intended ; Briefly resolved by means of certain Breviats.' *Quarto* (octavo size) (O).

There is also in the title-page ' Examined also and corrected at the Presse, by the Author himselfe.' A great many circumstances induce me to think that the general fashion of correcting the press by the author came in with the seventeenth century, or thereabouts. *Breviats* are tables. As far as I know, this is the first English book of tables of compound interest. And there are *real* tables of half-yearly and quarterly compound interest. Again, decimal fractions are really used: the tables being constructed for ten million of pounds, seven figures have to be cut off, and the reduction to shillings and pence with a *temporary* decimal separation, is introduced when wanted. For instance, when the quarterly table of amounts of interest at ten per cent is used for three years, the principal being 100*l.* (page 99), in the table stands 137266429, which multiplied by 100 and seven places cut off gives the first line of the following citation:

' The Worke

$$Facit \left\{ \begin{array}{rr|r} l & 1372 & 66429 \\ sh & 13 & 2858 \\ d & 3 & 4296.' \end{array} \right.$$

Giving 1372*l.* 13*s.* 3*d.* for the answer. And the tables are expressly stated to consist of *numerators*, with 100... for a denominator.

[A reprint of this work, by T. Fisher, *London*, sixteen-thirty-four, *duodecimo*, is in the Royal Society's Library.]

Paris, sixteen-fourteen. **Artabasda.** ' Nic. Smyrnæi Artabasdæ Græci Mathematici ΕΚΦΡΑΣΙC Numerorum Notationis per gestum digitorum. Item Venerab. **Bedæ** de Indigitatione et manuali loquelæ lib.' *Octavo in twos* (edited by F. Morell, the first Gr. Lat., the second Lat.).

These tracts are on nothing but the mode of representing numerals by the fingers. Bede's tract is not a separate work, but a chapter from his treatise *de natura rerum*, in which it will be found in the ' Bedæ . . . opuscula plura,' edited by Noviomagus, *Cologne*, fifteen-thirty-seven, *folio in twos.* But the first edition is, *Venice*, fifteen-twenty-five, Joh. Tacuinus (editor and printer), ' hoc in volumine hæc continentur M. Val. Probus . . . . . Ven. Beda de Computo per gestum digitorum . . . . nunc primum edita.' *Quarto in fours.*

Paris, sixteen-fourteen. **Honorat Meynier.** ' L'Arithmetique . . . . enrichie de ce que les plus doctes mathematiciens ont inventés . . . . tant aux formes que nos anciens ont practiquees comme en celles qui se practiquent aujourd'huy en France, en Holande, en Allemagne, en Espagne et autres nations.' *Quarto* (O).

There is a great deal of matter in this book, which runs to 664 pages; and it might be historically useful. Part of it is an attack on Stevinus, who, by the way, is represented as then alive. The following is one of the examples in subtraction: ' L'an 1535, Jean Calvin composa son labirinthe abominable (puisque dans iceluy s'est perdu un grand nombre d'hommes de bien) . . . je demande combien s'est passé d'années depuis qu'il composa le dit Dedale.'

London, sixteen-fourteen. **Thomas Bedwell.** ' De Numeris Geometricis. Of the nature and properties of geometrical numbers, first written by Lazarus Schonerus, and now Englished,' &c. *Quarto.*

On figurate, square, &c. numbers, with applications to mensuration.

Bologna, sixteen-sixteen. **P. A. Cataldi.** ' Quarta parte della pratica Aritmetica Dove si tratta della principalissima, et necessariissima Regola chiamata comunemente del Tre,' &c. *Folio* (quarto size).

A long and wordy treatise on the rule of three, but venturing on examples of a much higher order of difficulty of computation than had been previously attempted.

Edinburgh, sixteen-seventeen. **John Napier.** ' Rabdologiæ seu Numerationis per Virgulas Libri duo : cum Appendice de expeditissimo Multiplicationis Promptuario, Quibus accessit et Arithmeticæ Localis Liber unus. Authore et inventore Johanne Nepero,' &c. *Duodecimo* (O).

A posthumous work. Napier's rods are well known, and this book is the *descriptio princeps* of them, with applications. The use of decimal fractions, expressly attributed to Stevinus, renders it remarkable. It is stated (P. 441) that Napier invented the decimal point, but this is not correct: 1993·273 is written by him 19932'7''3'''

Oppenheym, sixteen-seventeen and nineteen. **Robert Fludd.** ' Utriusque Cosmi Majoris scilicet et minoris metaphysica, physica atque technica historia.' Two volumes. *Folio in twos.*

The second volume has another title-page, purporting that it gives the supernatural, natural, preternatural, and contranatural history. Of Robert Fludd (though he was great enough in his day to engage the attention of Descartes) I shall here say nothing, except that he gave his mixture of mysticism and science two dedications, one on each side of a leaf. The first, signed *Ego, homo*, was addressed to his Creator: the second, signed ' Robert Fludd' to James I. of England. The first volume contains a treatise on Arithmetic, and on *Cossic* Arithmetic, or Algebra. The arithmetic is rich in the description of numbers, the Boethian divisions of ratios, the musical system, and all that has any connexion with the numerical mysteries of the sixteenth century. The algebra is of the four rules only, referring for equations and other things to Stifel and Recorde. The signs of addition and subtraction are P and M with strokes drawn through them. The notation for powers is the *q, c, qq*, &c. series of symbols. Perhaps the most remarkable thing about the algebra is, that Fludd, who wrote it in France for the instruction of a Duke of Guise, should have known nothing of Vieta. The second volume is strong upon the hidden theological force of numbers.

London, sixteen-nineteen. **Henry Lyte.** 'The art of tens, or decimall arithmeticke.' *Octavo* (duodecimo size).

One of the earliest English users of decimal fractions. P. 440.

Hamburgi, sixteen-twenty. **Francis Brasser** and **Otto Wesellow.** 'Arithmetica Francisci Brasseri . . . . ab Ottone Wesellow ex Germanico in Latinum sermonem versa, atque in lucem edita.' *Octavo* (O).

Brasser seems to have been a celebrated teacher : it is not clear whether he published or not. The book itself is a mixture of arithmetic and algebra in Scheubel's form, with much of algebraical application to commercial questions. There is extraction of roots as far as the fourth.

Hamburg, sixteen-twenty-one. **Peter Lauremberg** of Rostoch. 'Institutiones Arithmeticæ.' *Octavo* (small).

A demonstrative book.

Rostoch, sixteen-twenty-one. **John Lauremberg.** 'Organum analogicum, sive instrumentum proportionum.' *Quarto.*

This instrument is the sector.

Rostoch, sixteen-twenty-three. **John Lauremberg.** 'Virgularum numeratricum et promptuarii arithmetici descriptio, figuræ et usus.' *Quarto.*

A tract of fourteen pages on Napier's rods.

London, sixteen-twenty-four. **William Ingpen.** 'The Secrets of numbers, according to theologicall, arithmeticall, geometricall, and Harmonicall computation.' *Quarto.*

A worthy follower of Peter Bungus. P. 425.

London, sixteen-twenty-four. **Thomas Clay.** 'Briefe easie and necessary tables of interest and rents forborne,' &c. *Octavo* (very small size).

There is no hint here of decimal fractions. Attached is a ' chorologicall discourse' on the management of estates.

London, sixteen-twenty-eight. **John Speidell.** 'An Arithmeticall Extraction or Collection of divers questions with their answers.' *Octavo* (small).

For John Speidell, see the article *Tables* in the Supplement of the *Penny Cyclopædia*. The book is literally what it professes to be, questions with their *facits* or answers, and nothing else, mostly in reduction, the rule of three, and practice.

Amsterdam, sixteen-twenty-nine. **Albert Girard.** 'Invention nouvelle en l'algebre,' &c. *Quarto*.

In this celebrated work there is a slight treatise on arithmetic, the most remarkable part of which is, that the author gives no examples in division by more than one figure, and seems to decline them as too difficult for his readers: he gives some results only. P. 426, 412. On one occasion he uses the decimal point.

Oughtred's *Clavis* is a book of arithmetic as well as of algebra, and one of great celebrity. The editions that have fallen in my way are:

London, sixteen-thirty-one. ' Arithmeticæ in numeris et speciebus institutio : quæ tum logisticæ, tum analyticæ, atque adeo totius mathematicæ, quasi clavis est.' Signed ' Guilelmus Oughtred,' at the end of the dedication. *Octavo*.

London, sixteen-forty-seven. 'The Key of the Mathematicks New Forged and Filed : Together with A Treatise of the Resolution of all kinde of Affected Æquations in Numbers. With the Rule of Compound Usury ; And demonstration of the Rule of false Position. And a most easie Art of delineating all manner of Plaine Sun-Dyalls. Geometrically taught by Will. Oughtred.' *Octavo*.

Oxford, sixteen-fifty-two. ' Guilelmi Oughtred . . . . . . Clavis Mathematicæ denuo limata, sive potius fabricata . . . . Editio tertia auctior et emendatior.' *Octavo*.

Oxford, sixteen-sixty-seven. ' Do. do. . . . . Editio quarta auctior et emendatior.' *Octavo*.

Oxford, sixteen-ninety-three. ' Do. do. . . . . Editio quinta auctior et emendatior.' *Octavo*.

Oxford, sixteen-ninety-eight. ' Do. do. Editio quinta auctior et emendatior. Ex recognitione D. Johannis Wallis, S.T.D. Geometriæ Professoris Saviliani.' *Octavo*.

London, sixteen-ninety-four (again in seventeen-two). 'Mr. William Oughtred's Key of the Mathematics. Newly translated . . . with Notes . . . . . . Recommended by Mr. E. Halley.' *Octavo*.

The third edition was an extended re-translation into Latin : and

E

the preface of it, which was copied into succeeding editions, shews that it was carefully performed under Oughtred's own eye. The treatise on dialling was translated into Latin by a young man of sixteen, of Wadham College, whom Oughtred describes as already an inventor in Astronomy, Gnomonics, Statics, and Mechanics, and of whose future fame he augurs great things : his name was Christopher Wren. There are two fifth editions in the list, the second of them revised, it appears, by Dr. Wallis, who ought, one would think, to have known better the history of the book he edited. But on examining them, I find that they are the same impressions with different title-pages : so that it would seem as if Wallis had allowed his eminent name to appear as a guarantee for a book he had never revised. Wrong again, however : for in the preface to the third edition it appears that Wallis had given all the help of an editor at that stage of the work ; so that his name should rather have appeared before.

The editions which follow the first, besides other additions, have the solution of adfected Equations, or Vieta's method (see the Index, Horner'). Throughout, the old, or *scratch* method of division, is retained. Dr. Peacock observes of this method that it lasted nearly to the end of the seventeenth century : but it thus appears that it even got into the eighteenth, and that with Halley's name as a recommendation. Throughout all the editions, Oughtred's ancient algebraical notation is retained, as also his way of writing decimal fractions (12|3456 for our 12·3456). I cannot tell why Dr. Peacock always spells his name Oughtred*e* (p. 441, &c.).

London, sixteen-thirty-two. **William Oughtred.** ' The Circles of Proportion and the Horizontal Instrument. Both invented and the uses of both Written in Latine by Mr. W. O. Translated into English : and set forth for the publique benefit by William Forster.' Published with it, *London,* sixteen-thirty-three, ' An addition unto the use of the instrument . . . . . For the Working of Nauticall Questions . . . . . Hereunto is also annexed the excellent Use of two Rulers for Calculation. And it is to follow after the 111 Page of the first Part.' *Quarto.*

The second edition, *Oxford,* sixteen-sixty, *octavo,* is ' by the Author's consent Revised, Corrected . . . . by A. H. Gent.' [The editor was Arthur Haughton.] The addition above named gives the first description of the *sliding rule.* The circles of proportion are what would now be called circular sliding rules; and the two rulers are the common sliding rule, the rulers being kept together by the hand. Various errors are afloat about the invention of the sliding rule. Some give it to Wingate, some to Gunter, some to Partridge; the official book of the Excise Office gives it to an excise officer whose name I forget. But the truth is that Oughtred invented it. (See Penny Cycl. ' Slide Rule.')

London, sixteen-thirty-three. **Robert Butler.** ' The

Scale of Interest; or Proportionall Tables and Breviats . . . . .
Together with the valuation of Annuities,' &c. *Octavo.*

These tables are for discount or present value; and as far as
annuities, &c. are concerned, that is, in compound-interest questions,
resemble Witt's. But decimals are really used in the applications,
the separating notation being as in 12|3456.

London, sixteen - thirty - three. **Nicholas Hunt, M.A.**
' The Hand-Maid to Arithmetick refined: Shewing the variety
and facility of working all Rules in whole Numbers and Fractions, after most pleasant and profitable waies. Abounding
with Tables above 150. for Monies, Measures and Weights, tale
and number of things here and in forraigne parts; verie use-
full for all Gentlemen, Captaines, Gunners, Shopkeepers, Arti-
ficers, and Negotiators of all sorts: Rules for Commutation
and Exchanges for Merchants and their Factors. A Table
from 1l. to 100 thousand, for proportionall expences, and to
reserve for Purchases.' *Octavo* (ON). P. 442.

The first thing that strikes any one in this book is the slavish-
ness of the dedication to an Earl, and the grotesque appearance
which the use of the ambiguous word *rare* gives it. " When I reflect on ancient Nobility, earth's glory, it being found in the way of
vertue; this so rare a thing transports my soule with thoughts of a
glorious eternitie. And for as much as the Lord hath said (*Dixit
dii estis*) that you are titular and tutelar terrestriall gods, powerfully
protecting the meaner shrubbes under the spreading branches of
your tall Cedars: I wonder not to see almost every Pamphlet to
crouch and deject it selfe in all humility and lowlinesse, seeking
patronage from persons of eminency." He afterwards calls on his
patron to " imitate the propitiousnesse of the divine essence," which
will excite him to " a voluntarie prostitution of his humble service,"
if the patron will protect him from " that squint-ey'd foole, that
antipathie to vertue (the envious man)."

The book itself is very full on weights and measures, and commercial matters generally. It does not treat of decimal fractions:
what the author calls ' decimall Arithmeticke' is the division of a
pound into 10 primes of two shillings each; each shilling into *six*
primes of two pence each. There are some verses, perhaps suggested by the republication of Buckley, headed: ' Arithmeticke-
Rithmeticall, or the Handmaid's Song of Numbers;' of which the
rules of addition and subtraction will give a sufficient specimen:

> Adde thou upright, reserving every tenne,
> And write the Digits downe all with thy pen,
>   The proofe (for truth I say,)
>   Is to cast nine away.
> From the particular summes, and severall
> Reject the Nines; likewise from the totall
> When figures like in both chance to remaine
> Clear light of working right shal be your gain;

Subtract the lesser from the great, noting the rest,
Or ten to borrow, you are ever prest,
To pay what borrowed was thinke it no paine,
But honesty redounding to your gaine.

Paris, sixteen-thirty-four (with another title-page, sixteen-forty-four, but it is all one impression). **Peter Herigone.** 'Cursus Mathematici tomus secundus.' *Octavo* (O).

The volumes of this course, which is said to be the first complete course published, have different dates. It is a polyglott book, Latin and French. This second volume contains practical arithmetic. It introduces the decimal fractions of Stevinus, having a chapter 'des nombres de la dixme.' The mark of the decimal is made by marking the place in which the last figure comes. Thus when 137 livres 16 sous is to be taken for 23 years 7 months, the product of 1378′ and 23583‴ is found to be 32497374″″ or 3249 liv. 14 sous, 8 deniers. There is a *memoria technica* for numbers in this book, which I subjoin, that those who know the old system of Grey may compare the two. The consonants and vowels which stand for numbers are :

P	B	C	D	T	F	G	L	M	N
1	2	3	4	5	6	7	8	9	0
a	e	i	o	u	ar	er	ir	or	ur $\rbrace$
					ra	re	ri	ro	ru

So that *r* signifies *five* except before or after *u*, when it destroys the meaning altogether. Grey's system is :

B	D	T	F	L	S	P	K	N	Z
1	2	3	4	5	6	7	8	9	0
a	e	i	o	u	au	oi	ei	ou	y

The vowel-systems of the two might be combined.

London, sixteen-thirty-four. **William Webster.** 'Webster's tables for simple interest direct . . . . Also his Tables for Compound Interest . . .' *Octavo* (small). Third edition.

This work treats decimal arithmetic as a thing known : but all the tables are not decimals. Neither is there yet any recognised decimal point ; only a partition-line to be used on occasion. In this book I find the first head-rule for turning a decimal fraction of a pound into shillings, pence, and farthings : though not so perfect a one as was afterwards found. I have seen the second edition, place, title, and form as above, sixteen-twenty-nine. The copy of each edition which I have examined is bound up with King James's list of customs, 'The Rates of Marchandizes' &c. promulgated in 1605.

London, sixteen-thirty-seven (again in forty-seven). **Leonard Digges.** 'A book named Tectonicon . . . . Published by Leonard Digges, Gentleman, in the yeere of our Lord 1556.' *Quarto*.

A book of mensuration : one of the instances in which a book was republished for the name of its author, long after its methods were obsolete. This occurred so frequently in the seventeenth century that we must take for granted the existence of a class of mathematicians to whom the light of Napier, &c. had not penetrated.

London, sixteen-thirty-eight ; reprinted in seventeen-forty-eight. **John Penkethman.** ' Artachthos, or A New Book declaring The Assise or Weight of Bread by Troy and Avoirdupois Weights. Containing divers Orders and Articles made and set forth by the Right Honble the Lords and Others of his Majesty's most Honble Privy Council . . . . . . Published by their Lordships Orders.' *Quarto.*

This does not look like a book of arithmetic; but Penkethman has thought it necessary to prefix instruction in numeration, both Roman and Arabic.

' Note that IV signifies IIII, as IX signifies nine which takes as it were by stealth, or pulls back one from foure and ten.' So that, in fact, I stands behind x and picks his pocket.

[*Leyden,* sixteen-forty.] **Adrian Metius.** ' Arithmetica Practica.' *Quarto* (N).

There is no prolixity about this book. Sexagesimal fractions are taught, but not decimal ones. The title is torn out in my copy.

Herbornæ Nassoviorum, sixteen-forty-one. **Joh. Henr. Alsted.** ' Methodus Admirandorum Mathematicorum Novem libris exhibens universam Mathesin.' *Duodecimo* (N).

This has a slight treatise on arithmetic, with a few words on algebra : nothing on decimal fractions.

Sora, sixteen-forty-three. **John Lauremberg.** ' Arithmetica . . . . exemplis historicis illustrata, itidem algebræ principia.' *Quarto* (N).

A book which uses the Italian method of division is a rarity among German books of this period. The examples are drawn from historical matters, and from the military art.

Paris, sixteen-forty-four. **Theon** of Smyrna. ' Τῶν κατὰ μαθηματικὴν χρησίμων εἰς τὴν τοῦ Πλάτωνος ἀνάγνωσιν.' *Quarto.* Edited by Ismael Bullialdus (Bouillaud).
Leyden, eighteen-twenty-seven. **Theon** of Smyrna. The same title. *Octavo in twos.* Edited by J. J. de Gelder.

These are the only editions of Theon of Smyrna, neither having the complete work, if indeed any complete manuscript exist. The first has the arithmetic and music; the second, with various readings, the arithmetic only. The arithmetic is nothing more than

E 2
3

that classification and nomenclature of numbers which is also given by Nicomachus (the two were contemporaries), and followed by Boethius. The music, as in other cases, is only discussion of fractions under names derived from the scale. Theon of Smyrna must not be confounded with his greater namesake of Alexandria.

London, sixteen-forty-five.   **Edmund Wingate.**   'Use of the Rule of Proportion in Arithmetique and Geometrie.' *Duodecimo.*   (A translation from Wingate's own French.)

This book is on the use of Gunter's logarithmic scale, not the sliding rule, which Wingate never wrote upon.

Leyden, sixteen-forty-six.   **Francis Vieta.**   'Opera Mathematica.'   *Folio in twos.*

In this collection is the ' De numerosa potestatum purarum atque adfectarum ad exegesin resolutione tractatus,' which was first published, *Paris*, sixteen-hundred, *folio*, and is the first exposition of the method by which the general evolution of the roots of equations is connected in operation and principle with the extraction of the square root and common division. See the references presently given under the name of Horner. The first departure from Vieta's form of solution, which amounts to incorporating with it the method afterwards known by the name of Newton, was made by Henry Briggs in the 'Trigonometria Britannica,' *Gouda*, sixteen-thirty-three, *folio in twos and threes.* Accordingly, Briggs has a fair claim to be the first user of that which is called Newton's method of approximation, though he does not explain it, but leaves the explanation to be collected from his examples. The work which introduced Vieta's method into England was the 'Artis Analyticæ Praxis' of Thomas Harriot, *London*, sixteen-thirty-one, *folio.* All these writers would have called their process algebraical, but the progress of the art of computation will make their works essential parts of the history of arithmetic. I believe the most complete account of Vieta is that which I have given in the *Penny Cyclopædia :* with which may be read the account of his works in Hutton's *Tracts*, vol. ii. pp. 260-274.

London, sixteen-forty-eight.   **Seth Partridge.**   'Rabdologia, or the Art of numbring by Rods.'   *Duodecimo* (O).

The use of Napier's rods or bones explained.

Napier's *Rabdologia* brought the well-known *Napier's rods* into vogue for half a century (P. 411). And the reason seems to have been that this contrivance really was found useful, in the state of the habit of computation as it then existed. P. 432, 440.

Leyden, sixteen-forty-nine.   **Joh. Hen. Alsted.**   'Scientiarum omnium Encyclopædia.'   Two volumes.   *Folio in threes.*

The first edition is said to be sixteen-thirty, and the preface of this one is dated sixteen-twenty-nine, but it refers to a course of philosophy published in sixteen-twenty, of which it seems to be an enlargement. As far as I can learn, this is the first of the works to which moderns attach the idea of a *Cyclopædia*. For though Ringelberg and Martinius had published under that name, and though the *Margarita Philosophica* above mentioned and other works ran through the whole circle of elementary knowledge — yet I cannot establish that any work before Alsted's was so large and comprehensive as to be a Cyclopædia in the sense in which Chamber's Dictionary was afterwards so called. The course of arithmetic (O) in the second volume is rather scanty, but seems less so when the bareness of the professed mathematical courses is considered. There is a table of the squares and cubes of all numbers up to 1000. Logarithms are not mentioned, but Napier's rods are. Moreover, by the date of the first edition, we must call this a course of mathematics prior to Herigone's, usually called the first.

London, sixteen-fifty. **John Wybard.** 'Tactometria. seu, Tetagmenometria. Or, The Geometry of Regulars practically proposed .... by J. W.' *Octavo in twos.*

An excellent book of mensuration of solids, full of remarkable information on the subject of weights and measures.

London, sixteen-fifty. **Jonas Moore.** 'Moore's Arithmetick : Discovering the secrets of that Art, in Numbers and Species.' *Octavo* (N).

This is Jonas Moore's first work, and is a very good one. It is very complete in decimals, giving the contracted multiplication and division. It has the use of logarithms (in which the operations with the negative characteristic are fully given), algebra from Oughtred, and squares and cubes of all numbers up to 1000, and the fourth, fifth, and sixth powers up to 200. The second edition is :

London, sixteen-sixty. 'Moor's Arithmetick in two books.' *Octavo.*

It is worthy of remark that the common arithmetic of this edition is in black letter, the second book, or Algebra, being Roman letter, as all the previous edition was. Perhaps Recorde, &c. had given common readers a prejudice for black letter arithmetics. P. 442.

Dantzig, sixteen-fifty-two. **Joh. Broscius.** 'Apologia pro Aristotele et Euclide . . . . . Additæ sunt duæ disceptationes de numeris perfectis.' *Quarto.*

A perfect number is one which is equal to the sum of its less factors : thus 28, being the sum of 1, 2, 4, 7, and 14, is a perfect number. And a number was called defective or redundant, according as the sum of its factors fell short of or exceeded the number. Thus 8 is defective, and 12 is redundant.

London, sixteen-fifty-three. **Noah Bridges.** 'Vulgar Arithmetique, explayning the Secrets of that Art, after a more exact and easie way than ever.' *Octavo* (N).

London, sixteen-sixty-one. **Noah Bridges.** ' Lux Mercatoria, Arithmetick Natural and Decimal,' &c. *Octavo* (N).

Dr. Peacock (p. 452) has given the first book notoriety by his citations from the eulogistic poetry at the beginning, which he has by no means exhausted. The poem of Geo. Wharton, the royalist astrologer, on the existing state of things, is a political satire. Another commends the plainness of the work, hints at decimals and logarithms becoming too common, and pronounces that a merchant

> ' may fetch home the Indies, and not know
> What Napier could, or what Oughtred can doe.'

Dr. Peacock speaks lightly of the work, in mentioning the excessive praises with which it was announced. But a very short treatise, explicit upon the modern mode of division, which those who praised it had perhaps never seen before, and upon the use of practice, was rather an exception to the rule at the time at which this book was published.

The second work is more learned, and has an appendix on decimals. The author disapproves of the use which some would make of decimals, and avers that the rule of practice is more convenient in many cases; which is perfectly true. Bridges seems to have had a very clear view of the capabilities of arithmetic without formal fractions. He mentions some preceding authors, and his list is 'The Merchant's Jewel,' 1628; 'The Handmaid to Arithmetick,' 1633; 'The Map of Commerce,' 1638; Masterman, Johnson, Hill, Leybourn, Recorde, Baker, Val. Menher, Boteler.

London, [sixteen-fifty-four]. **E[dmund] W[ingate].** ' Ludus Mathematicus. Or, The mathematical Game : Explaining the description, construction,, and use of the Numericall Table of Proportion.' *Duodecimo.* (Again in eighty-one.)

This is the description of a logarithmic instrument, which it would be impossible to give a notion of without the instrument itself, or a drawing. The date is cut out in my copy.

London, sixteen-fifty-five. **Thos. Gibson.** ' Syntaxis Mathematica.' *Octavo.*

This work is filled with the construction of equations from Des Cartes; but it gives the arithmetical solution, or Vieta's Exegesis, from Harriot: also interest-tables, partly from *Simon Stevens.*

Oxonii, sixteen-fifty-seven. **John Wallis.** ' Mathesis Universalis : sive Arithmeticum Opus Integrum,' &c. *Quarto.*

This and other works, with their separate title-pages, are collected (apparently so published) as *Opera Mathematica*, and were afterwards republished, in substance, in the folio collection of Wallis's works. The Arithmetic gives both modes of division : but the old notation, as 12|345, is used for decimal fractions. Wallis afterwards adopted the decimal point in his Algebra.

Wurtzburg, sixteen-fifty-seven-and-eight. **Gaspar Schott.** ' Magia Universalis Naturæ et Artis.' *Quarto* (four parts, variously bound in volumes).

This work, and others of a similar character by the same author, are the precursors of the works of mathematical amusement compiled by Ozanam, Montucla, Hutton, &c. In the fourth part he treats of arithmetic, mostly on the wonders of combinations. The gem of the book is his accurate calculation of the degrees of grace and glory of the Virgin Mary, which are exactly

$$115792089237316195423570985008687907853269984665640564039457584007913129639936$$

not one more nor less. This is the 256th power of 2, and is repeated three times, printed in the same way. Others solved the same problem, and found out the number of stars at the same time, by writing down every possible way in which the words of

Tot tibi sunt dofes, Virgo, quot sidera cœlo

can be arranged in an hexameter line.

As Schott here describes the digit of his geometrical pace as being forty poppy-seeds placed side by side, I thought it possible I might get a fellow to the barley measure (see page 8). But it seems that he has not played fair, or has thought it necessary to bring the degrees of glory of the Virgin up to the 256th power of 2, at any sacrifice. For I find that forty poppy-seeds, side by side, measure an inch and a half (English) ; so that the geometrical foot of sixteen digits would be two feet English.

Leyden, sixteen-sixty. **Vincent Leotaud.** ' Institutionum arithmeticarum libri quatuor, in quibus omnia quæ ad numeros simplices, fractos, radicales, et proportionales pertinent precepta clarissimis demonstrationibus tum arithmeticis tum geometricis illustrata traduntur.' *Quarto* (O).

This is a clear but heavy work, running to 700 pages.

Herbipoli(Wurtzburg), sixteen-sixty-one. **Gaspar Schott.** ' Cursus Mathematicus.' *Folio in threes.*

It has a shabby arithmetic, which, considering the magnitude of the course, reminds us of Falstaff's halfpennyworth of bread. In the frontispiece, a lion and a bear draw a car surmounted by an armillary sphere, and having for its wheels (or rather castors) the celestial and terrestrial globes, along an amphitheatre the pavement of which is studded with diagrams. These diagrams contain a sly little bit of freemasonry. Schott, as a good Jesuit, adheres to the Ptolemaic system, grudgingly, and because his church had condemned the other, as appears from his words. But in the diagrams on the pavement of his frontispiece, he draws the Copernican system, and no other. Dr. Peacock (p. 408) refers to the 'Arithmetica Practica' of Schott [*Herb.* sixteen-sixty-two——].

London, sixteen-sixty-four. **James Hodder.** 'Hodder's Arithmetick : or, That necessary Art made most easie.' The third edition, much enlarged. *Duodecimo* (O).

Had this work given the new mode of division, it must have stood in the place of Cocker. The ninth edition was published in sixteen-seventy-two. [The first edition is said to have had the date 1661.] In Davis's sale catalogue (1686), there are marked as by Hodder, a Vulgar Arithmetic, 1681, and a Decimal Arithmetic, 1671. There are no decimal fractions in the book before us.

Deventer, sixteen-sixty-seven. **Joachim Camerarius.** 'Explicatio in duos libros Nicomachi Geraseni .... et notæ **Samuelis Tennulii** in Arithmeticam Jamblichi ....' *Quarto.* Also, (one publication, though different in place and date,)

Arnhem, sixteen-sixty-eight. **Jamblichus.** 'In Nicomachi Geraseni Arithmeticam introductionem et de Fato.' *Quarto* (Gr. Lat.).

These comments of Jamblichus and Camerarius are the easiest accesses to a knowledge of the matter of the work of Nicomachus, of whom no Latin version exists, of which I can find any mention.

London, sixteen-sixty-eight. **John Newton.** 'The Scale of Interest : Or the Use of Decimal Fractions.' *Octavo.*

This book is further described in 'Tables,' *Penny Cyclop. Suppl.* : it was expressly intended for a school-book, though it is a strange one for the time.

London, sixteen-sixty-eight. **William Leybourn.**

A

Platform { for } Purchasers
Guide { for } Builders
Mate { for } Measurers.

Octavo (O).

The first book is on interest, the second and third on building and mensuration of work.

Toulouse, sixteen-seventy. **Diophantus.** 'Arithmeticorum libri sex.' *Folio in twos.*

This is perhaps the best edition of Diophantus. It has the com-

mentaries of Bachet and of Fermat. The character of the work of
Diophantus on the properties of numbers, treated algebraically, is
too well known to need description. I do not consider him as belonging to the genuine Greek school. His date is always spoken
of as very uncertain, though the preponderance of opinion seems
to place him in the second century. But I think I have given
sufficient reason for supposing him to have written as late as the
beginning of the seventh century. (See Dr. Smith's Dictionary of
Biography, article *Hypsicles*.)

Diophantus is usually placed in the second century, because
Suidas says that the celebrated Hypatia wrote a comment on the
astronomical table of *Diophantus*. This was a common name : Fabricius has collected upwards of twenty writers or philosophers who
bore it; and the arithmetician of whom I am speaking shews no
appearance* of having attended to astronomy. Diophantus mentions Hypsicles, a mathematician : now (unless this mention make
a second) there is only one Hypsicles in Greek literature, namely,
the author of the two last books of Euclid's elements. Of Hypsicles, Suidas says that he was a pupil of " Isidore the philosopher."
But in another place (and this passage, occurring at the word *Syri-
anus*, has been overlooked till now) Suidas quotes Damascius for his
information upon this Isidore. Now it is certain that the Isidore,
whose life John Damascius wrote, must have been a contemporary of
the Emperor Justinian, at the earliest: and his pupil, Hypsicles, can
hardly have written before the middle of the sixth century. Diophantus, who makes mention of Hypsicles, may have flourished
towards the end of the sixth, or the beginning of the seventh century. The silence of Proclus, Pappus, and Theon, as to both Hypsicles and Diophantus, is a strong presumption of the latter two
having come after the former three. When Dr. Peacock (p. 397)
says that Theon was well acquainted with the writings of Diophantus, I suppose it is upon the presumption that the father and
teacher could not have been ignorant of the author on whom the
daughter and pupil was said to have published a commentary. But,
this apart, I cannot imagine any reason to suppose that Theon had
ever seen the work of Diophantus. P. 404.

London, sixteen-seventy-three. **Sam. Morland.** 'The
Description and Use of two Arithmetick Instruments, together
With a Short Treatise explaining and Demonstrating the Ordinary Operations of Arithmetick, As likewise A Perpetual Almanac And several Useful Tables.' *Octavo* (duodecimo size).

There is also the following second title-page (first in order of time)
which is all that some copies have : *London*, sixteen-seventy-two.
'A New, and most useful Instrument for Addition, and Subtraction of Pounds, Shillings, Pence, and Farthings. . . . By S. Morland.'

---

* For this reason Fabricius wants to alter the words " a commentary upon of Diophantus the astronomical table," into " a commentary upon Diophantus and an astronomical table."

A very miscellaneous book, embodying computation (N), some of Euclid, tables for Easter, description of the calculating machine, &c.

London, sixteen-seventy-three. **John Kersey.** ' Mr Wingate's Arithmetick, containing a plain and familiar method,' &c. *Octavo.*

This is called the sixth edition of Wingate (N), one of the very best of the old writers. This edition has an appendix in augmentation of the commercial part, by Kersey, who says the first edition (which I have never seen) was published about sixteen-twenty-nine.

Amsterdam, sixteen-seventy-three. **William Bartjen.** ' Verniende cyfferinge . . . .' *Octavo* (small) (O).

A book of a decidedly commercial character, and with good force of examples.

London, sixteen-seventy-four. **John Mayne.** ' Socius Mercatoris: or the merchant's companion, in three parts.' *Octavo* (N).

The three parts are on arithmetic, vulgar and decimal, on interest, and on solids and cask-gauging: the second and third have two separate title-pages, dated sixteen-seventy-three. There is a little algebra.

London, sixteen-seventy-four. **John Collins.** ' An Introduction to Merchants-Accompts.' *Folio in twos.*

Book-keeping by double entry. Collins says that his first edition was in 1652, and his second (which was nearly all destroyed by the fire) in 1664-5.

London, sixteen-seventy-six. **A. Forbes.** ' The Whole Body of Arithmetick Made Easie.' *Duodecimo* (O).

Not a very easy system.

London, sixteen-seventy-seven. **Michael Dary.** ' Interest epitomised, both compound and simple . . . . whereunto is added, a Short Appendix For the Solution of Adfected Equations in Numbers by Approachment: Performed by Logarithms.' *Quarto* (octavo size) (N).

Of Dary, who was a gunner, then a tobacco-cutter, then a teacher, &c., and who was the correspondent of Collins, J. Gregory, Newton, &c., a good deal is to be found in the Macclesfield correspondence. This tract contains the distinct announcement and use of a principle which is now well known, and of which Newton's (or Briggs's) method of approximation is only a particular case. If an equation be reduced to the form $x=\phi x$, ϕx being a known function of x, successive formation of $\phi a=b$, $\phi b=c$, $\phi c=d$, &c., gives an approximation to a root of the equation, whenever the results do not increase without limit. Dary saw and applied this, expediting the process occasionally by using the mean of the two last results, instead of the last result alone, to produce the next result. The book itself is quite out of notice; and neither the ingenuity nor the importance of the principle was appreciated by Dary's contemporaries.

London, sixteen-seventy-eight. **William Leybourn.** 'Arithmetick Vulgar, Decimal, Instrumental, Algebraical.' *Octavo.* Fourth edition. (The four parts have separate titles and paging.) (ON).

I have not met with any earlier edition. This work seems to have been substantially inserted in the course of mathematics by Leybourn, presently mentioned.

London, sixteen-eighty. **Thomas Lawson.** 'A Mite into the Treasury, being a Word to Artists, Especially to Heptatechnists.' *Quarto.*

This book I have several times met with in lists, as one of arithmetic, so that I here insert a true account of it. It is written against human learning, and arithmetic among the rest, as being of no use to "open the seals of the book," or to interpret the Bible. It reminds me much, difference of age and manners apart, of some books I have lately seen, of arithmetical examples, in which no name is introduced except from the Bible : so that Judas Iscariot and the devil may be mentioned, but Socrates or Milton may not. Of arithmetic T. Lawson says: "Herein any member of Italian Babilon with his Mass-Book, Mass for the Dead, Fabulous Legend : any Mahumetan with his dreggy Alcoran ; any Flint-hearted Jew with his Talmud, a mingle-mangle of Jewish Divine and Humane matters; any Dead, Dry, Unfruitful Formalist, may grow Profound, Exquisit, Nimble, yea, and though involved in the intricate windings of Degeneration, out of the Royal state of Regeneration and Heavenly Transformation may apprehend the Feates, Termes and Parts of this Natural Art, as Digits, Articles mixt Numbers, Cyphers, Terniries, Golden Rule direct, Golden rule reverse, a Cube, Phythagoras's Table, Algorism, &c. yet be Strangers to the Divine Exercise which leads to Christ, the Lion of the Tribe of Judah, who alone opens the Seals of the Book." Attention is then directed to the number of the Beast.

London, sixteen-eighty-one. **Jonas Moore.** 'A new Systeme of the Mathematicks.' Two volumes. *Quarto.*

The first volume contains arithmetic, much abridged from Moore's larger work above noted. This course was written for Christ's Hospital.

London, sixteen-eighty-two. **Gilbert Clark.** 'Ough-

F

tredus explicatus, sive Commentarius in Ejus Clavem Mathematicam.' *Octavo.*

This commentary is now of no use. A wise commentator will endeavour to fill his book as much as he can with such little matters as will be historically interesting a couple of centuries after his own time; and this will in many cases cause him to outlive his author. I suppose this commentator is the Gilb. *Clerke* who published, in sixteen-eighty-seven, on what he called the *Spot-dial.*

London, sixteen-eighty-three. **Peter Galtruchius.** 'Mathematicæ totius . . . Clara, Brevis et accurata Institutio.' *Octavo.*

There is a short treatise on arithmetic (O), and logarithms are mentioned as a recent invention, without any examples of their use.

London, sixteen-eighty-four. **Rich. Dafforne.** 'The Merchant's Mirrour.' *Folio in twos.*

A book on book-keeping, translated from the Dutch.

London, sixteen-eighty-four. **Abr. Liset.** 'Amphithalami or the Accountants closet.' *Folio.*

Book-keeping by double entry. This and the two last books on the subject (Dafforne and Collins) I have bound together with the *Lex Mercatoria* of Gerard Malynes and other books. Mr. M'Culloch has precisely the same collection, and gives them all as one publication (*Liter. Polit. Econ.* p. 129-30). But they have different printers, different dates, and different signatures: most being in twos, and the rest simple folios ; and from Collins's preface it seems that his book was certainly a separate publication. I merely mention this (having seen the like in other cases) to advert to the probability of it not having been unusual for booksellers to bind parcels of books on the same subject together for sale.

London, sixteen-eighty-four. **Wm. Leybourn.** 'The line of Proportion or [of ?] Numbers, Commonly called Gunter's Line made easie,' &c. *Duodecimo.*

On the sliding rule, which invention is attributed to Seth Partridge. But he says that Milbourn (see Sherburne's Manilius, or *Penny Cycl.* 'Horrocks') 'disposed it in a Serpentine or Spiral Line ;' the logarithmic spiral, of course (*Penn. Cycl.* 'Slide Rule').

London, sixteen-eighty-four. **Thos. Baker.** 'The Geometrical Key : or the Gate of Equations unlocked :' &c. *Quarto.*

Polyglott : Latin and English. An attempt to do without arithmetic in the solution of equations, by means of construction. See the correspondence of Baker and Collins in the Macclesfield correspondence, beginning of vol. ii.

London, sixteen-eighty-five. **Seth Partridge.** 'The description and use of an instrument called the double scale of proportion.' *Octavo*.

On the sliding rule.

London, sixteen-eighty-five. **John Collins.** 'The doctrine of decimal arithmetick, simple interest, as also of compound interest . . . . made publick by T. D.' *Octavo* (small).

A clear and short work, posthumous.

Naples, sixteen-eighty-seven. **Ægid. Franc. de Gottignies.** 'Logistica Universalis.' *Folio in twos*.

The author is the man mentioned by Montucla (ii. 643) as having claimed some of the discoveries of Cassini. He was a Jesuit, and was professor at Rome. This work is the last of several, of which it seems to contain the substance, amplified (Heilbronner, *in verb.*). I have never met with this book till very lately; and, though I have had no time as yet to examine it fully, I have been much surprised to find it unnoticed by mathematical historians. In his views of algebra, the author seems much before his time. His recognition of its existence at that time as an art only; of its containing principles and definitions which are not arithmetic, but perfectly distinct; its definition as a science in which subtraction is universalised by express convention; the avowed enlargement of meaning of + and − &c. &c. — shew that the author had got more than a glimpse of what was coming. The book is a folio of 400 pages, with much prolixity of expression, but vigorous efforts of thought.

London, sixteen-eighty-seven. **William Hunt.** 'The Gauger's Magazine.'

A treatise on decimal fractions and mensuration, once well known. There is a table of squares in it up to that of 10,000.

London, sixteen-eighty-seven. **D. Abercromby.** 'Academia Scientiarum. . . . . Being a Short & Easie Introduction to the Knowledge Of the Liberal Arts and Sciences.' *Octavo* (small).

A real smatterer's book, published contemporaneously with the *Principia* of Newton. It gives names and general notions, with the names of a few celebrated authors; just enough to set up a man about·town. There is another book, which may go with this, out of its place, being the earliest I have met with of the kind.

London, sixteen-forty-eight. **Sir Balthazar Gerbier.** 'The interpreter of the Academie for forrain languages, and all noble sciences, and exercises.' *Quarto*.

A polyglott, English and French, announcing the intention of setting up a College in London. It gives general ideas on the sciences in dialogue ; and that which relates to arithmetic passes between the tutor and a page of the court, who begins thus : " If you would favour me now with some particular discourses on all the necessarie sciences (which would make me capable of the Theoricall part), I should esteeme myselfe very much obliged unto you for that it would make me passe for a gallant wit in good companies of men that love to discourse pertinently of things." With this view, my predecessor (in intention) gives two pages and a half of arithmetical terms, and then the young gentleman says : " Ther's enough for my Theory." This book was dated from Paris the day before the funeral of Charles I. of England: but it was dedicated to the Duke of York, with a hope that his turn for literature might increase the comfort of his great parents.

London, sixteen-eighty-eight. **Jonas Moore.** ' Moore's Arithmetick : in Four Books.' The third edition. *Octavo.*

A complete book of Arithmetic (N), with book III. on Logarithms, and book IV. on algebra, both dated sixteen-eighty-seven. This book was edited by John Hawkins, Cocker's editor, who has put his name to the last three parts. See more of this presently, under *Cocker.*

London, sixteen-eighty-eight. **Peter Halliman.** ' The Square and Cube Root compleated and made easie.' *Octavo* (second edition) (O).

That this book should have reached a second edition shews how great an object it was to save a little calculation. The author simply approximates to the fractional part of the root by first interpolation. Thus his general formula would be :

$$\sqrt[n]{(a^n + b)} = a + \frac{b}{(a+1)^n - a^n}$$

to help which he has a small table of squares and cubes, with first differences. But he evidently seems to think he has got the *exact* square and cube roots : and he says :

> Now Logarithms lowre your sail,
> And Algebra give place,
> For here is found, that ne'er doth fail,
> A nearer way, to your disgrace.

London, sixteen-ninety. **Henry Coggeshall.** ' The Art of Practical Measuring Easily Performed, By a Two Foot Rule Which slides to a Foot,' &c. *Duodecimo in threes.*

On the sliding rule and the use of logarithms.

London, sixteen-ninety. **William Leybourn.** ' Cursus Mathematicus . . . in nine books.' *Folio in twos and fours.*

The arithmetic (ON) Vulgar, Decimal, Instrumental, and Alge-braical, is stated to be the first book : but algebra is afterwards pro-moted to follow geometry. Leybourn gives both methods of division, but thinks the old one not inferior to the new.

Leyden, sixteen-ninety. **Claude Fr. Milliet Dechales.** ' Cursus seu Mundus Mathematicus.' *Folio in twos.* Four volumes.

The first volume contains a course of Arithmetic (O), on which I may remark, as just now in the case of Schott, that it is very scanty. It is, perhaps, only in a collected course of Mathematics, that we see how difficult a thing computation must have been considered. And I begin to see perfectly, by instance after instance, that it was not the Greeks only who fled to geometry to escape from arithmetic.

Amsterdam, sixteen-ninety-two. **Bernard Lamy.** ' Ele-mens des Mathematiques,' &c. *Duodecimo.* Third edition (O).

The arithmetic is very poor : p. 410, a friend sends him the mode of using the *carriage* in subtraction ; he having previously *borrowed* from the upper line : he prints this as a novelty.

London, sixteen-ninety-three. **John Wing.** ' Heptar-chia Mathematica . . . . Arithmatick . . . .' *Octavo* (N).

A plain book, containing, among other things, arithmetic, and a larger collection than usual of tables useful in mensuration. It has the licenser's permission dated after the Revolution (Sept. 14, 1689).

Paris, sixteen-ninety-three. **Nicolas Frenicle.** In the ' Divers Ouvrages de Mathématique et de Physique,' *folio,* are contained the following tracts of Frenicle, namely, ' Methode pour trouver la solution des Problemes par les exclusions— Abregé des combinaisons—Des Quarrez magiques—Table ge-nerale des Quarrez magiques de quatre de costé.' And in another collection of Frenicle's tracts only, *quarto,* no date nor place, with only fly title-pages (pp. 374, register A—Yy), there are the preceding tracts and also ' Traité des Triangles Rect-angles en nombres,' in two parts, [of which last there is said to be an edition *Paris,* sixteen-seventy-six, *octavo,* which may be the one just mentioned, for aught I know].

Frenicle (who died in 1675) was one of those *useless* men, as many appliers of mathematics have called them, who work subjects into shape before the time is come for applying them. In this point we have degenerated from our ancestors in the soundness of our views as to what knowledge is, and how it comes. For instance, at the end of the seventeenth century, Picard, an eminently useful (as well as *useful*) applier of mathematics, who never speculated on any subject, was the careful preserver of Frenicle's manuscript combinations, and exclusions, and magic squares. A century after,

Condorcet, a useful (but *useless*) speculator, who never handled any so-called *practical* subject, finds it necessary, in his *éloge* of Frenicle, to apologise for the *useless* character of his writings, though the doctrine of probabilities,—our power over which depends on the knowledge of combinations, &c., in which Frenicle was a successful workman—was, when the *éloge* was written, almost as far advanced as the theory of probable errors and the method of least squares, on which every observer who knows his business now relies, and to which astronomy in particular is much indebted for its modern accuracy. Condorcet had caught a slang which was current in his day, and has been very popular among us. The time is coming when really learned men will again be ashamed of not seeing the value of all the uses of mind : when nothing but thoughtlessness or impudence, mercurial brain or brazen forehead, will aver that no knowledge is practical, except that which ends in the use of material instruments.

Paris, sixteen-ninety-four.   **Jean Prestet.**   ' Nouveaux Elemens de Mathematiques.'   *Quarto* (two volumes, third edition) (ON).

This work is a complete body of arithmetic and algebra, treated with depth and clearness, and embracing a great extent of subject. The author is much more impressed with the necessity of strict and sustained reasoning in algebra than most of his contemporaries ; and in particular, his demonstrations of the processes of arithmetic are ample; a very uncommon thing in his day. Montucla, who does not appear to have known this third edition (for he mentions only the two former ones as of sixteen-seventy-five and eighty-nine), dismisses it with an opinion that it is *ouvrage très estimable*. He speaks of many inferior works with much higher praise.

London, sixteen-ninety-four.   **Wm. Leybourn.**   ' Pleasure with Profit : Consisting of Recreations of Divers Kinds, viz. Numerical, &c.'   *Folio in twos*.

The recreations on arithmetic contain a great many of those which are still common. But some have gone out of vogue. I do not remember ever having had the pleasure or profit of the following :—To put together five odd numbers to make 20. Answered as follows :—Three nines turned upside down, and two units. Honest Leybourn thinks this answer is a *Falacy*; in which I differ from him : I think the question more than answered, viz. in *very* odd numbers.

London (no date, but about sixteen-ninety-five).   **Venterus Mandey.**   ' Synopsis Mathematica Universalis : or the Universal Mathematical Synopsis of John James Heinlin, Prelate of Bebenhusan.'   *Octavo*.

Translated from the third edition, Latin, *Tubingen*, sixteen-ninety. Heinlin's system of arithmetic (O) is somewhat theological. ' In Unity all numbers are virtually, although they are infinite. So all things are in God, in him they live and move. Seven is a Sacred Number, chiefly used in Holy Scripture. It seems to have its Original from the Inscrutable Unity of Divine Essence, and Sextuple in respect of the Divine Persons among themselves: whence also the Description of a Septangled form is impossible, and cannot be known by Human Minds.'

London, sixteen-ninety-six. **Samuel Jeake.** 'Λογιστικη-λογία, or Arithmetick Surveighed and Reviewed.' *Folio* (ON).

To see the size and weight of this book, one would have thought arithmetic had been a branch of controversial divinity. But it is now very valuable, from the variety of the information which it contains, particularly on weights, measures, and coins; and from the goodness of the index. There is a good deal of algebra in it, many quaint names, and stories of the same kind. Thus, in stating the famous story of the Delian oracle telling the inquirers to double a cube, in order to rid themselves of the plague, it is stated that the meaning was, that the best method to deliver realms from such contagion, was to abate of their voluptuousness, and apply themselves to literature. But for all this, those who know the value of a large book with a good index, will pick this one up when they can. P. 442.

Paris, sixteen-ninety-seven. **De Lagny.** 'Nouveaux Elemens d'Arithmetique et d'Algebre.' *Octavo* (duodecimo size).

He has the new method of division, but is obliged to write down the divisor afresh every time he wants to use it. As an instance of the effect of change of times and methods, the following trivial circumstance may be worth repeating. De Lagny says that it would take an ordinary computer a month to find the integers in the cube root of 696536483318640035073641037. To shew what would be thought of such an assertion now, I had put down the work for all the integers and a few decimals, for insertion in the *Companion to the Almanac.* When the time came for making up the manuscript, the slip of paper on which the above was written found itself (as the French say, which was more than I could do) in a heap of unarranged papers: and it was better worth my while to repeat De Lagny's month's work, than to sort the papers and pick it out.

Oxford, sixteen-ninety-eight. **E. Wells.** 'Elementa arithmeticæ numerosæ et speciosæ.' *Quarto* (octavo size) (N).

This is an elegant work, in which arithmetic and algebra are exhibited to the university student in a simple form, accompanied by the historical learning which such a student ought to have. The works on arithmetic were taking too commercial a turn to be all-sufficient for the purposes of liberal education : and Wells, without by any means rejecting commercial questions, writes in order that the student may not of necessity be driven to the works in which, as he says, *exempla non aliunde petuntur quam a Butyro et Caseo, Zingibere et Pipere, aliisque consimilibus.*

London, no date (but printed for W. and J. Marshal). **Andrew Tacquet.** ' The Elements of Arithmetick In Three Books, The Seventh, Eighth and Ninth of Euclid : With the Practical Arithmetic In Two Books . . . . Translated into English by an Eminent Hand.' *Octavo.*

Dr. Peacock mentions Tacquet as a late instance of the old method of division. Both methods are given in this translation, "whereof the first," meaning the Italian method, "is least used but is the best; the other most used, but the difficultest." [According to Dr. Peacock, the original is " Arithmeticæ Theoria et Praxis," *Antwerp*, sixteen-fifty-six ——.] I have described an edition of Tacquet elsewhere (see the Index).

London, seventeen-hundred. **Edward Cocker.** 'Cocker's Arithmetick : Being A plain and familiar Method, suitable to the meanest Capacity for the full understanding of that Incomparable Art, as it is now taught by the ablest School-Masters in City and Country. Composed By Edward Cocker, late Practitioner in the Arts of Writing Arithmetick, and Engraving. Being that so long since promised to the World. Perused and Published by John Hawkins Writing-Master near St. Georges Church in Southwark, by the Authors correct Copy, and commended to the World by many eminent Mathematicians and Writing-Masters in and near London.' *Duodecimo.*

This is called "The *twentieth* edition carefully corrected, with additions:" but I find that it had become usual, when any one edition was augmented, to make the assertion with reference to all succeeding editions. The first edition of this famous work was in sixteen-seventy-seven (I have seen one copy, which appeared in a sale a few years ago), the fourth in sixteen-eighty-two, the twentieth as above, the thirty-third in seventeen-fifteen, the thirty-fifth in seventeen-eighteen, the thirty-seventh in seventeen-twenty. The earliest edition I ever possessed is one of sixteen-eighty-five; what edition it is, is not stated. But there is confusion among the title-pages. For though the above is unmistakeably marked seventeen-hundred, and twentieth edition, I have also compared together two editions, both of seventeen-*twelve*, one called the *fourteenth*, and the other the *thirtieth* : all three by one printer, Eben. Tracy, at the three Bibles on London Bridge. How long it went on in England I do not know : but there was an edition at Edinburgh in seventeen-sixty-five, and another at Glasgow in seventeen-seventy-one, both edited by John Mair, whose preface is dated seventeen-fifty-one. Some account of Cocker is given in the *Penny Cyclopædia* (art.

' Cocker'). Not much more is known of him than this, that he was a skilful writing-master and engraver. At the beginning of the Arithmetic is a recommendation signed " John Collens" (no doubt the famous John Collins, or intended to pass for him) certifying that the deceased author was " knowing and studious in the Mysteries of Numbers and Algebra, of which he had some choice Manuscripts and a great collection of Printed Authors in several Languages." Collins doubts not "but he hath writ his Arithmetick suitable to his own Preface and worthy acceptation:" which means that Collins or Collens had only seen the preface of the forthcoming work at most. Then follows the attestation of fifteen teachers to the merits of the work. All this looks odd; because, according to the editor, the book was that which had been long promised to the world by a celebrated writer. All attestation was unnecessary; and the certificate of a celebrated name, wrong spelt, to the effect that he had no doubt the work, then printed, would be good, may now excite a little curiosity, if not suspicion.

I am perfectly satisfied that Cocker's Arithmetic is a forgery of Hawkins, with some assistance, it may be, from Cocker's papers: that is to say, there has certainly been more or less of forgery, without any evidence being left as to whether it was more or less. I could easily believe that all was forged; and my reasons are as follows:

In both the editions of Hodder which I have seen (1664 and 1672) is the following advertisement: " There is newly printed Mr. Cocker's book called the Tutor to Writing and Arithmetic." It appears then that during his lifetime he had published a book on Arithmetic, which I suspect to have been what would now be called an arithmetical copy-book, with engraved questions and space left for the work. But neither the posthumous work, nor its preface signed by Cocker himself, make the least allusion either to the previous work, or to the promise of another. On the contrary, the language of Cocker's own preface implies that it is the first work he has published on arithmetic, and agrees with many other prefaces (which are usually written last) in speaking of the work *as already published.* To establish these and other contradictions, I first give Hawkins's account in *his* preface (with my own Italics). " I Having the Happiness of an Intimate Acquaintance with Mr. Cocker in his Life-time often solicited him to remember his Promise to the World, of Publishing his Arithmetick, *but* (for Reasons best known to himself) *he refused it;* and (after his Death) the Copy falling *accidentally* into my hands, I thought it not convenient to smother a work of so considerable a moment," &c. But Cocker himself writes, or is made to write, as follows: " By the sacred Influence of Divine Providence, I have been Instrumental to the benefit of many; by virtue of those useful Arts, Writing and Engraving: and do *now* with the same *wonted alacrity* cast this my Arithmetical Mite into the Publick Treasury. . . . For you the pretended Numerists* . . .

---

* *Numerist* is a word which Hawkins uses in his own professed writings: and it was by no means a common word.

was this Book composed and published. . . ." This is an odd preface for a book which the author never meant to publish, and refused to publish, though pressed to do so. Of course it is possible
that though he wrote with an intention of publishing, he afterwards
changed his mind. This is one explanation; that Hawkins forged
clumsily is another: and which is the most probable must be
gathered from a review of all the circumstances.

Next, at the end of the work, Hawkins gives a hint of a book on
decimals which would be forthcoming in time. Accordingly, we
have:

London, sixteen-eighty-five. ' Cocker's Decimal Arithmetick . . .
Whereunto is added his Artificial Arithmetick . . . . also his Alge-
braical Arithmetick . . . . according to the Method used by Mr.
John Kersey in his Incomparable Treatise of Algebra. Composed
by Edw. Cocker . . . . Perused, Corrected, and Published by John
Hawkins . . . .' *Octavo*.

This book came to its third edition in seventeen-two. The artificial
(or logarithmic) arithmetic, and the algebra, have separate title-
pages, dated sixteen-eighty-four. Cocker gives no preface here,
but Hawkins does, stating that he had in the preceding work given
" an account of the speedy publication of his Decimal, Logarithmical,
and Algebraical Arithmetick." He has here mended his hand: for,
except the words " such Questions being more applicable to Decimals are omitted till we come to acquaint the Learner therewith,"
the first treatise does not give a hint of the second. Again, Kersey's
Algebra, on which part of Cocker's second work is founded, was
published in 1673, and the latter had been dead some time before
the manuscript of the first work of 1677 " accidentally" fell into
Hawkins's hands. This is again singular, on any supposition but
that of the forgery. Moreover, at the end of the preface, Hawkins
writes a letter to his friend John Perkes, *in cipher* (*Penny Cycl.*
' Cocker') in which he says, " If you peleas to bestow some of your
spare houres in perusing the following tereatise, you will then be
the better able to judg how I have spent mine." This looks like a
confession of authorship. And in 1704, as presently noted, appeared
Cocker's English Dictionary by *John Hawkins*, who would perhaps,
had he lived, have found Cocker's Complete Dancing-Master and
Cookery-book among the papers of the deceased.

The famous book itself I take to be a compilation or close imitation in all its parts. Even the Frontispiece, &c. is fashioned upon
Hodder. Thus, Hodder begins with his own portrait, and verses
of exaggerated praise under it; and so does Cocker. The former
begins his title with *Hodder's Arithmetick*, the latter with *Cocker's
Arithmetic*. The former speaks of that *necessary art*, the latter of
that *incomparable Art*. The former has it ' explained in a way
familiar to the *capacity*, the latter a '*familiar* method suitable to
the meanest *capacity*:' the two words being then by no means so
common in the senses put upon them as they are now. Turning
over the title-page we find that each of them ' humbly dedicateth
this Manual (manuel in Hodder) of Arithmetick,' the first to a

most worthily honoured friend,' the second to 'much honoured friends;' the first 'in token of true gratitude for *unmerited* kindnesses,' the second ' as an acknowledgment of *unmerited* favours.' There are too many small coincidences here.   And it must be remembered that every resemblance to a work so well known as Hodder's (it is one of the few English works of the century which have found their way into Heilbronner's list) would help the sale.

From all these circumstances I was tolerably sure that there was no dependence to be placed on the famous Cocker being any body but Hawkins, so far as this book is concerned : though I must say I hardly expected to find such confirmation as would arise from catching Hawkins at a similar trick in another quarter.   But on looking at the work above described as the third edition of Jonas Moore's Arithmetick, my eye was caught by the following sentence : " You may likewise prove Division by Division, as I have shewed at large in the 7. chap.   Page 100, 101, 102. of Mr. Cocker's Arithmetic, printed in the year 1685."   Now Jonas Moore was dead before 1685 ; and moreover, could have shewn nothing in Cocker's Arithmetic : and on looking farther to see who it is that thus speaks in the first person, I find the name of John Hawkins to the second part of the work, as editor.   And on looking farther, I find that a good deal from Moore's *own* editions has been introduced *verbatim* into Cocker.   For instance, this sentence is in both : " Notation teacheth how to describe any number by certain notes and [or] characters, and to declare the value thereof being so described."   And throughout the book, paragraphs are frequently introduced from Moore, with alterations of phrase here and there. So that we have Hawkins arbitrarily altering and adding, in the first person, to the text of a book which had been for thirty years before the world under Moore's name.   What are we to suppose he would do with Cocker's papers, if indeed he had any ?   Moreover, we find in Cocker sentences which had been previously written by Moore.

To see whether much was gained by Cocker's Arithmetic, as well as for the interest of the comparison itself, I will write down the definitions of addition, subtraction, multiplication, and division, from Recorde, Wingate, Johnson (see Additions to this work), Moore, Bridges, Hodder, and Cocker.

ADDITION.

Recorde. Addition is the reduction and bringing of two summes or more into one.

Wingate. Addition is that by which divers Numbers are added together, to the end that their sum, aggregate, or total, may be discovered.

Johnson. Addition serveth to adde or collect divers summes of severall denominations, and to expresse their totall value in one summe.

Moore. Addition is that part of Numbring or Numeration,

whereby two or more numbers are added together, and so the totall
or summe of them is formed.

Bridges. Addition is the gathering together and bringing of
two numbers or more into one summe.

Hodder. Addition teacheth you to add two or more sums together, to make them one whole or total sum.

Cocker. Addition is the Reduction of two, or more numbers of
like kind together into one Sum or Total.   Or it is that by which
divers numbers are added together, to the end that the Sum or Total
value of them all may be discovered.

SUBTRACTION.

Recorde. Subtraction diminisheth a grosse sum by withdrawing
of other from it, so that Subtraction or Rebating is nothing els, but
an arte to withdrawe and abate one sum from another, that the Re-
mainer may appeare.

Wingate. Subtraction is that by which one number is taken out
of another, to the end that the remainder, or difference, between the
two numbers given may be known.

Johnson. Subtraction serveth to deduct one summe from another; the lesser from the greater, and to shew the remaines.

Moore. Substraction is that part of Numeration where one number is substracted or taken out of another, and so the Remainder is
gotten, which is also called the difference or excesse.

Bridges. Subtraction is the taking of one number from another,
whereby the residue, remainder or difference is found.

Hodder. Substraction teacheth to take any lesser number out of
a greater, and to know what remains.

Cocker. Subtraction is the taking of a lesser number out of a
greater of like kind, whereby to find out a third number, being or
declaring the Inequality, excess, or difference between the numbers
given, or Subtraction is that by which one number is taken out of
another number given, to the end that the residue, or remainder
may be known, which remainder is also called the rest, Remainder,
or difference of the numbers given.

MULTIPLICATION.

Recorde. Multiplication is such an operation, that by two summes
producyth the thirde : whiche thirde summe so manye tymes shall
containe the fyrst, as there are unites in the second.   And it serveth in the steede of many Additions.

Wingate. Multiplication teacheth how by two numbers given to
find a third, which shall contain either of the numbers given, so
many times as the other contains 1 or unitie.

Johnson. Multiplication is a number of additions speedily performed.

Moore. Multiplication, is a part of conjunct Numeration, or numbring, whereby the Multiplicand (which is the number to be multiplied) is so often added to it selfe, as an unite is contained in the

Multiplyer (which is the number multiplying) and so the *Factus* (or Product) which is the result of the worke, is had.

Bridges. Multiplication (which serveth for many additions) is that by which we multiply two numbers the one by the other, to the end their product may be discovered.

Hodder. Multiplication serveth instead of many additions, and teacheth of two numbers given to increase the greater as often as there are Unites in the lesser.

Cocker. Multiplication is performed by two numbers of like kind, for the production of a third, which shall have such reason to the one, as the other hath to unite, and in effect is a most breif and artificial compound Addition of many equal numbers of like kind into one sum. Or Multiplication is that by which we multiply two or more numbers, the one into the other, to the end that their Product may come forth, or be discovered. Or, Multiplication is the increasing of any one number by another; so often as there are Units in that number, by which the other is increased, or by having two numbers given to find a third, which shall contain one of the numbers as many times as there are Units in the other.

DIVISION.

Recorde. Division is a partition of a greater summe by a lesser.

Wingate. Division is that by which we discover how often one number is contained in another, or (which is the same) it sheweth how to divide a number propounded into as many equal parts as you please.

Johnson. Apparently considers division not enough of a technical term to need definition : his first example is, " I would divide 65490 pound amongst 5 men."

Moore. Division is that part of conjunct Numeration, wherby one Number is substracted from another, as often as it is contained in it, and by that meanes it is found how many of the one is contained in the other.

Bridges. Division is that by which we discover how often one number is contained in another.

Hodder. Division is that by which we know how many times a lesser sum is contained in a greater.

Cocker. Division is the Separation, or Parting of any Number, or Quantity given, into any parts assigned ; Or to find how often one Number is Contained in another; Or from any two Numbers given to find a third that shall consist of so many Units, as the one of those two given Numbers is Comprehended or contained in the other.

The six predecessors of Cocker whom I have chosen, stop when they think enough has been said. But the illustrious discoverer, or at least the first general propagator, of the fact that two and two make four (for his current reputation amounts to this) must have had more in view. He seems to be laying every offence against accuracy in different ways, so that the unfortunate schoolboy who commits it may be sure of a flogging under one count or another of

G

his definition. And the vice of confounding abstract and concrete number, which leads him to imply that five shillings can be multiplied by five shillings, runs through his whole book : as does also the tendency to prolixity and reduplication of things which confuse each other. As to the general notion of what Arithmetic is, Cocker tells his beginners that it is either Natural, Artificial, Analytical, Algebraical, Lineal, or Instrumental. The natural is "that which is performed by the Numbers themselves; and this is either Positive or Negative. Positive, which is wrought by certain infallible numbers propounded, and this either Single or Comparative; Single, which considereth the nature of numbers simply by themselves; and Comparative, which is wrought by numbers as they have Relation one to another. And the Negative part relates to the Rule of False." Artificial Arithmetic is performed by artificial or borrowed numbers invented for that purpose, called Logarithms. Analytical Arithmetic " is that which shews from a thing unknown to find truly that which is sought; always keeping the Species without Change." Algebraical Arithmetic " is an obscure and hidden Art of accompting by numbers in resolving of hard Questions." Lineal Arithmetic "is that which is performed by lines, fitted to proportions, as also Geometrical projections." Instrumental Arithmetic " is that which is Performed by Instruments, fitted with Circular and Right lines of proportions, by the motion of an index or otherwise." So much for Cocker (or Hawkins) as an explainer. As to the actual modes of operation, they are neither better nor worse than those pointed out before by Wingate, Moore, and Bridges. The famous book looks like a patchwork collection, and, I believe, is nothing more. The reason of its reputation I take to be the intrinsic goodness of the processes, in which the book has nothing original; and the systematic puffing with which it was introduced. The long-promised book of the great Mr. Cocker, with Collins and fifteen other teachers to recommend it, pushed aside better productions. I am of opinion that a very great deterioriation in elementary works on arithmetic is to be traced from the time at which the book called after Cocker began to prevail. This same Edward Cocker must have had great reputation, since a bad book under his name pushed out the good ones.

London, seventeen-hundred. **Christopher Sturmius.** ' Mathesis Enucleata, or the Elements of the Mathematicks. . . . . Made English by J. R. M. and R. S. S.' *Octavo.*

Arithmetic and Algebra are here very nearly separated.

Edinburgh, seventeen-one. **George Brown.** ' A Compendious, but a Compleat System of Decimal Arithmetick, Containing more Exact Rules for ordering Infinites, than any hitherto extant . . . . . First Course.' *Quarto.*

The author's knowledge of what was then extant, seems far from complete.

London, seventeen-three. **John Parsons** and **Thos. Wastell.** ' Clavis Arithmeticæ.' *Octavo* (small) (ON).

The old system of division is rather recommended. There is a very neat work on algebra at the end.

Amsterdam, seventeen-four. **Andrew Tacquet.** 'Arithmeticæ Theoria et Praxis.' *Octavo* (small) (ON).

There had been several preceding editions. The theory consists in a version of the seventh, eighth, and ninth books of Euclid : the practice in an ordinary treatise. Both methods of division are given : the Italian as best, but least used.

London, seventeen-six. **William Jones.** ' Synopsis Palmariorum Matheseos ; or a New Introduction to the Mathematics.' *Octavo in twos* (N).

Jones is well known among the contemporaries of Newton, and was the father of his celebrated namesake, the Indian Judge. The book, as its name imports, is a kind of syllabus.

London, seventeen-seven. ****, Art. Bac. Trin. Col. Dub. ' Arithmetica absque Algebra aut Euclide demonstrata. Cui accesserunt, Cogitata nonnulla de Radicibus Surdis, de Æstu Aeris, de Ludo Algebraico, &c.' *Octavo in twos* (NO).

There is no doubt that the author was the celebrated Bishop Berkeley, then a youth under twenty. The object of the arithmetic is stated in the title ; and at that time the effort was much wanted. The algebraical game defies brief explanation.

London, seventeen-seven. **John Smart.** 'Tables of simple interest and discount.' *Octavo in twos* (small).

This is the first edition of these celebrated tables, and is little known. The second edition, *London,* seventeen-twenty-six, *quarto,* is the best having Smart's name. But in reality the Tables in Francis Baily's 'Doctrine of Interest and Annuities,' *London,* eighteen-eight, *quarto,* are Smart's, and do not profess to be any thing else. Mr. Baily (who, by the way, did not know this first edition) says, ' I have neither time nor inclination to calculate them anew ; and therefore I give them to the world with all their imperfections on their head. I am happy, however, to observe, that after many years experience, I have not met with any errors but such as might be discovered on inspection.' Brand's edition of seventeen-eighty is said by Mr. Baily to have a good many errors.

There is yet another edition, *London,* seventeen-thirty-six. John Smart. ' Tables of Interest, &c. Abridged for the Use of Schools, in Order to Instruct Young Gentlemen in Decimal Fractions.' *Octavo.* Here is another instance of what I have before remarked, that compound interest has always been considered *the* application of decimal fractions, by those whose arithmetic has been commercial.

Manuscript (in my possession) no place, after seventeen-ten.

I insert this here, as proving that, so late as the date above mentioned, there were French schools in which the decimal point was not introduced, the old method of division was employed, and the Ptolemaic system was taught. It is the collection of notes of lectures made by a young Englishman educated in France, and was sold a few years ago by his descendant.

London, seventeen-fourteen. **Samuel Cunn.** 'A New and Compleat Treatise of the Doctrine of Fractions, Vulgar and Decimal,' &c. *Octavo in twos* (small) (N).

Prefixed is a testimonial from Halley, vouching for the goodness of the work and the novelty of some of its rules.

London, seventeen-fourteen. **Edward Wells.** 'The young Gentleman's Course of Mathematicks.' Three volumes. *Octavo*.

These volumes (in my copy) exhibit what I have more often found at the beginning of the eighteenth century than at any other time; namely, volumes of different editions in one set. The Arithmetic title-page of the first volume, which follows the general title-page, has 'seventeen-twenty-three,' and 'second edition.' The other volumes have seventeen-fourteen and seventeen-eighteen. This Wells is the same whose work I have mentioned above; he qualifies himself D.D. and rector of Cotesbach in Leicestershire. When I quoted his diatribe against butter and cheese, ginger and pepper (which I did before I had seen this work), I sympathised with him, thinking he meant that liberal education had its wants as well as professional. But I was mistaken: it is *gentlemanly* education, as opposed to that of "the meaner part of mankind," that he wants to provide for. Every page is headed on one side 'The young gentleman's,' on the other 'arithmetic,' 'geometry,' or 'mechanics,' as the case may be. The gentlemen are those whom God has relieved from the necessity of working, for which he expects they should exercise the faculties of their minds to his greater glory. But they must not 'be so Brisk and Airy, as to think, that the knowing how to cast Accompt is requisite only for such Underlings as Shop-keepers or Trades-men;' and, for the sake of taking care of themselves, 'no Gentleman ought to think Arithmetick below Him, that do's not think an Estate below Him.' This Wells might be made as useful now as the Spartans used to make their slaves. The Arithmetic is an abridged version of the work of sixteen-ninety-eight above described.

London, seventeen-fourteen. **Joh. Ayres.** 'Arithmetick Made Easie For the Use and Benefit of Trades-Men.' *Duodecimo*. Twelfth edition; with an Appendix on Book-keeping by Chas. Snell.

A work of the immediate school of Cocker.

London, seventeen-fifteen. **John Hawkins.** 'Cocker's English Dictionary.' Second edition. *Octavo in twos* (small).

I have entered this book here only because Hawkins asserts that it was the work of the celebrated arithmetician : which I do not believe, for the reason above given. [The first edition is said to have the date seventeen-four.]

London, seventeen-seventeen. **Wm. Hawney.** 'The Compleat Measurer . . . .' *Duodecimo in threes.*

A full treatise on decimal arithmetic.

No place marked, seventeen-$^{\text{seventeen}}_{\text{eighteen}}$. **George Brown.** 'Arithmetica Infinita, or the Accurate Accomptant's Best Companion.' *Octavo in twos* (small, oblong).

This is not a work on arithmetic, but a set of tables, which will certainly be reprinted as soon as the decimals of a pound gain their proper footing. The main part of it is the first nine multiples (and the 365th) of the decimals which express each farthing of the pound. Thus under 4*s,* 1¼*d.* are given the multiples of 20520833..... The whole work is copperplate engraving from beginning to end. From several indications, I gather that Geo. Brown of this work is also the author mentioned under seventeen-one.

London, seventeen-seventeen. **Roger Rea.** 'The Sector and Plain Scale, Compared . . . . Unto which is annexed, So much of Decimal Arithmatick and the Extraction of the square Root, as is necessary for the Working of Arithmetical Trigonometry.' *Octavo* (N).

The treatise of an illiterate and confused person. Nothing has been more common than for those who write on application to consider it advisable not to trust the books to teach, nor the readers to know, decimal fractions, and to supply a fresh treatise. Rea says he uses the Italian mode of division (N) as being that which is *most commonly used :* nothing more than this, even in 1717.

London, seventeen-eighteen. **Wm. Bridges.** 'An Essay to facilitate Vulgar Fractions ; After a New Method, and to make Arithmetical Operations Very Concise :' &c. *Duodecimo.*

London, seventeen-nineteen. **Good.** 'Measuring made Easy.' *Octavo* (duodecimo size).

A description of Coggeshall's sliding-rule,. corrected and enlarged by James Atkinson.

London, seventeen-nineteen. **John Ward.** 'The Young Mathematician's Guide.' The third edition. *Octavo in twos* (N).

This useful course, which commences with arithmetic [was first published about seventeen-six]. It is recommended by Raphson and Ditton. The sixth edition was in seventeen-thirty-four; the eighth edition was in seventeen-forty-seven.

London, seventeen-twenty-one. **Wm. Beverege** (Bp. of St. Asaph). ' Institutionum Chronologicarum Libri Duo. Unà cum totidem Arithmetices Chronologicæ Libellis.' Third edition. *Octavo in fours.*

The date of the preface is sixteen-sixty-eight. The arithmetical part is a treatise on the numerals of different nations, learned, but not always judicious, according to modern views of the history of symbols. It is followed by a brief elementary treatise on arithmetic, with chronological examples.

London, seventeen-twenty-six. **E. Hatton.** 'The Merchant's Magazine : or, Trades-Man's Treasury.' *Quarto.*

This is the eighth edition of a work of some celebrity, but which must not be confounded with Hatton's edition of Recorde. The only guide to the date of the first edition which I have is the statement of the eighth that it was reviewed in the *Ouvrages des Savans** for 1695.

It is interspersed with copperplate pages of flourished writing, containing examples and definitions. There is somewhat more of reason given for rules than was very common, and a vast quantity of mercantile terms, usages, &c. are explained.

Witemberg, seventeen-twenty-seven. **Joh. Fr. Weidler.** ' De Characteribus Numerorum Vulgaribus et eorum Ætatibus . . . . Dissertatio Critico-mathematica.' *Quarto.*

London, seventeen-twenty-eight. **E. Hatton.** ' A Mathematical Manual : or Delightful Associate.' *Octavo.*

Mostly on the use of the globes, but containing some " mysterious curiosities in numbers."

London, seventeen-thirty. **Alexander Malcolm.** ' A New System of Arithmetick, Theorical and Practical.' *Quarto.*

One of the most extensive and erudite books of the last century, having 640 heavy quarto pages of small type; " wherein," to go on with the title-page, " the science of numbers is demonstrated in a regular course from its first principles through all the parts and branches thereof, either known to the ancients or owing to the improvements of the moderns; the practice and application to the affairs of life and commerce being also fully explained : so as to

---

* Or else the *Acta Eruditorum.* This comes of translating. The phrase is " works of the learned."

make the whole a complete system of theory, for the purposes of men of science ; and of practice for men of business.   I quote this lengthy title as a true description of the work, at the date of publication.   Probably the union of such masses of scientific and commercial arithmetic made the book unusable for either purpose.

London, seventeen-thirty-one. **Edw. Hatton.** ' An Intire System of Arithmetic . . . . containing, I. Vulgar. II. Decimal. III. Duodecimal. IV. Sexagesimal. V. Political. VI. Logarithmical. VII. Lineal. VIII. Instrumental. IX. Algebraical.   With the Arithmetic of Negatives and Approximation or [sic] Converging Series,' &c.   Second edition.   *Quarto.*

A sound, elaborate, unreadable work, of 500 pages, of the same character as Malcolm's.

London, seventeen-thirty-one. **Wm. Hodgkin.**   ' A Short New and Easy Method of Working the Rule of Practice in Arithmetick.'   *Octavo in twos.*

An author who chooses his own examples can write a short method on any rule : but the first example taken at hazard will probably defy the abbreviations.

London, seventeen-thirty-two. **Joseph Champion.** ' Practical Arithmetick compleat.'   *Octavo in twos.*

London, seventeen-thirty-five. **John Kirkby.** ' Arithmetical Institutions, containing a Compleat System of Arithmetic, Natural, Logarithmical, and Algebraical.'   *Quarto in ones.*

A system of arithmetic is mixed up with algebra.   In the extraction of roots, Halley's formula is applied in such a manner as to make the operation seem continuous, though it is just as difficult as before.

London, seventeen-thirty-five. **James Lostau.** ' The Manual Mercantile, Second Book : Concerning Decimal Arithmetic . . . . .'   *Quarto.*

The first book was never published.   This work contains a slight treatise on Arithmetic, but the body of it consists of all the various integers and fractions that may be useful in commerce, with the first nine multiples of each.   It is 452 pages entirely of copperplate, the figures being rudely worked in, apparently by the author's own hand.   It is a posthumous work, and the editor says it took 17 years.   This mode of stereotyping was adopted in several instances in the first half of the last century.   And it must be observed, that if decimal arithmetic have not thriven in commercial affairs, it has not been for want of a great many attempts to facilitate the use of it, by publishing books of multiples.

London, seventeen-thirty-five. **Benjamin Martin.** 'A new Compleat and Universal System or Body of Decimal Arithmetick.' *Octavo in twos.*

A very full system of decimal arithmetic, applied to all parts of commercial arithmetic.

London, seventeen-thirty-six. **Thomas Weston.** 'A New and Compendious Treatise of Arithmetick.' *Quarto.* Second edition.

A simple and useful treatise.

Edinburgh, seventeen-thirty-six. ———. 'Arithmeticæ et Algebræ Compendium.' *Octavo in twos* (N).

There is a small treatise on arithmetic. The publishers are Thos. and Wal. Ruddiman.

London, seventeen-thirty-eight. **William Pardon.** 'A New and Compendious System of Practical Arithmetick.' *Octavo in twos.*

Four hundred full octavo pages is not a very compendious book on arithmetic, or would not be so now : but it looked small by the side of Hatton, Malcolm, and Kirkby.

London, seventeen-thirty-eight. **Tho. Everard.** 'Stereometry . . . . by the Help of a Sliding-Rule.' Edited by Leadbetter, tenth edition. *Duodecimo.*

A book which once had a great reputation among excise collectors.

London, seventeen-thirty-nine. **Christian Wolff.** 'A treatise of Algebra,' translated from the Latin by J. H. M. A. *Octavo.*

There is very little of arithmetic, and that mostly on the properties of numbers.

Cambridge, seventeen-forty. **Nicholas Saunderson.** 'The Elements of Algebra, in ten books.' *Quarto,* two volumes.

This is a posthumous work of the well-known Professor Saunderson, the blind lecturer on optics. The first volume contains a synopsis of Arithmetic, and the editor's account of the calculating board, by which Saunderson supplied the want of sight.

London, seventeen-forty. **Wm. Webster.** 'Arithmetick in Epitome.' Sixth edition. *Duodecimo* (ON).

The eighth edition of the author's book-keeping is *London,* seventeen-forty-*four*, *octavo in twos* (small); and the eighth edition of his 'Attempt towards rendering the Education of Youth more easy and effectual,' is *London*, seventeen-forty-*three* (with paging continued from that of the last). He says, "When a Man has tried all Shifts, and still failed, if he can but scratch out any thing like a fair *Character*, tho' never so stiff and unnatural, and has got but *Arithmetick* enough in his Head to compute the Minutes in a Year, or the Inches in a Mile, he makes his last Recourse to a Garret, and, with the Painter's Help, sets up for a Teacher of *Writing* and *Arithmetick ;* where, by the Bait of low Prices, he perhaps gathers a Number of Scholars.''

London, seventeen-forty. —— 'A small Treatise of the Square and Cube . . . .' Second edition. *Quarto* (O). Also,

London, seventeen-forty. 'A Supplement to the Square and Cube . . . .' *Quarto in twos.*

The author of this treatise (the last, I think, in which I have seen the old method of extracting the square root) is a copier of Peter Halliman, or some similar authority (see sixteen-eighty-eight), only his denominator is less by a unit.

London, seventeen-forty. **Rob. Shirtcliffe.** 'The Theory and Practice of Gauging,' &c.

A work once held in high estimation by the revenue-officers.

Edinburgh, seventeen-forty-one. **John Wilson.** 'An introduction to Arithmetick.' *Octavo in twos.*

A good demonstrative book, in a large type; very full on the *complete* operations with circulating decimals, the *ignes fatui* which have led many an arithmetical writer astray.

London, seventeen-forty-two. **John Marsh.** 'Decimal Arithmetic made perfect.' *Quarto.*

Almost entirely on infinite or circulating decimals. The predecessors whom he cites in his history of the subject are Wallis; Jones, 1706; Ward; Brown, 1708 or 1709 (he has not the work, but it is above at seventeen-one); Malcolm; Cunn; Wright, 1734; Martin, 1735; and Pardon. This subject of circulating decimals was at one time suffered to embarrass books of practical arithmetic, which need have no more to do with them than books on mensuration with the complete quadrature of the circle.

Leipzig, seventeen-forty-two. **Jo. Christoph. Heilbronner.** 'Historia Matheseos Universæ.' *Quarto.*

Though called a history of mathematics, and really a bibliography *raisonée*, yet it is peculiarly devoted to arithmetic, the authors on which have a separate list. There are also dissertations on numerals, on their history, &c. The index of this book is of rare goodness.

Geneva, seventeen-forty-three, forty-six, forty-seven, forty-nine, forty-one. **Christian Wolff.** 'Elementa Matheseos Universæ.' *Quarto.* A second or later edition (N). Five volumes.

The first of the five volumes contains a short treatise on arithmetic. Here, as happens so often in works of this period, a set is made up out of different editions. The first four volumes have *editio novissima*, the fifth has only *nova*. Wolff's course would be better known if it were scarcer. The ordinary reader passes it by as an old book; the collector as one which is very common. But it is replete with pieces of information, which are historical references and suggestions. As far as I can remember, Wolff is much the most learned historian of those who have written extensive courses.

London, seventeen-forty-five. **John Hill.** 'Arithmetick, Both in the Theory and Practice.' *Octavo in twos.*

This is the seventh edition of a work of much celebrity. It seems to have owed its fame partly to a recommendation of Humphrey Ditton, prefixed to the first edition (about 1712), praising it in the strongest terms. Perhaps at this time the only things which would catch the eye are the table of logarithms at the end, and the powers of 2 up to the 144th, very useful for laying up grains of corn on the squares of a chess-board, ruining people by horseshoe bargains, and other approved problems.

London, seventeen-forty-eight. **Charles Leadbetter.** 'The Young Mathematician's Companion.' *Duodecimo in threes.*

Second edition. Begins with an ordinary treatise on arithmetic.

London, seventeen-forty-eight. **William Halfpenny.** 'Arithmetick and Measurement, Improved by Examples.' *Octavo in twos* (N).

This is a surveyor's and artisan's book of application: but it contains decimal fractions.

London, seventeen-forty-nine; second part, seventeen-forty-eight. **Solomon Lowe.** 'Arithmetic in two parts. *Duodecimo in threes.*

This is a work both learned and foolish: but with the learning and folly so distinct that they can be used separately. The folly consists mostly in an attempt to give the rules of Arithmetic in English hexameters and in alphabetical order; I give a couple of instances.

Barter.

Barter, exchange of commodities: the rule to proportion 'em as follows:
What's to be changd, Value: then, see what That will purchase of T'other.
If an advanc'd price of one, a proportionable find for the other.

Casting out of Nines.

Prove by a careful review: 'tis the safest: the readiest, as follows:
Sub.] right; when -hend and remainder (together) make up the compound.
Add Mult Div] add the digits together and cast-out the nines : then
Right ; if remainder of Facits agrees with remainder of factors,
Multiplied in Mul: -sor and quotient in Div; to which add the remainder.

The learning consists in a great knowledge of former writers,
and a copious account of weights, measures, and coins: together
with a list of authors, which I have copied into my own, so far as
I could not find the names elsewhere.

London, seventeen-fifty. **James Dodson.** 'The Accountant, or the method of Book-keeping Deduced from Clear
Principles.' *Quarto.*

As far as I can find, this is the first book in which double
entry is applied to retail trade. James Dodson (my great-grand-
father) is best known to mathematicians in general by his *Antiloga-
rithmic Canon, London,* seventeen-forty-two, *folio* (see *Penny Cycl.*
'Tables').

London, seventeen-fifty. **Daniel Fenning.** 'The Young
Algebraist's Companion.' *Duodecimo in threes.*

There is a system of fractional arithmetic in this book, which
is written in dialogue. The author thought it impossible to understand algebra without some better works on arithmetical fractions
than then existed. As it is, says he, it is impossible to understand
the *Algorithm* much less the *Algorism,* which he explains by saying
that the former means the first principles, and the latter their practice. In this curious confusion of terms we see at its commencement
an instance of a process which is always going on (though in this
instance it has been arrested), the attachment of different meanings
to different spellings of the same word. My curiosity led me to take
a little trouble to trace Fenning to his authorities. And I find that
of two writers who must have been in his hands, Saunderson and
Kirkby, the first uses *Algorithm* for first principles, and the second
Algorism for practical rules. I think I remember having seen a
comparatively recent edition of this work.

London, ———. **R. T. Heath,** assisted by **W. Davidson.** 'The Practical Arithmetician: or Art of Numbers improved.' *Duodecimo.* (Revised by J. Bettesworth.)

Robert Heath was a person who made noise in his day, and in so
doing established a claim to be considered a worthless vagabond.
He was editor of the *Ladies' Diary* from 1746 to 1753, when the
Stationers' Company found it absolutely necessary to strike out some
of his scurrility, and dismiss him; appointing Thomas Simpson in

his place. But before this, in 1749, Heath had commenced the *Palladium*, an annual publication resembling the *Ladies' Diary*, the very first mathematical question of which is so expressed as to convey an indecent double-meaning, in a manner obviously intended. From 1750 or thereabouts, he began to write against Thomas Simpson. One of his publications against the latter is headed, " Miss Billingsgate in a salivation for a black eye:" and in a letter published in a newspaper, in 1751, he remarks of Simpson and another, that " the best writer against both is one who shall sign the warrant for their execution."

London, seventeen-fifty-one. **T. Smith.** ' Compendious Division. Containing, A Great Variety of Curious and Easy Contractions of Division.' *Octavo in twos*.

London, seventeen-fifty-three. **Sam. Stonehouse.** ' A Compendious Treatise of Arithmetic, By Way of Question and Answer.' *Octavo*. Third edition.

Question and answer is well when the difficulty is in the question and the solution in the answer: but to turn " Decimals are divided like integers" into " Are decimals divided like integers? The division is exactly the same," is trifling. This book gives the abbreviated rule for decimals of a pound, which very few books have given.

London, seventeen-fifty-six. **John Playford.** ' Vade Mecum, or the Necessary Pocket Companion.' Nineteenth edition. *Quarto* (long and narrow size).

This book is a ready reckoner, with miscellaneous tables. I have no information about the origin of these books, which I think are not so ancient as many may suppose. I could almost think from the preface (but such deductions are very deceptive) that the earliest of the books which are now called ready reckoners, meaning those which have totals at given prices ready cast up, was the following: *London*, sixteen-ninety-three. Wm. Leybourn. ' Panarithmologia; Being A Mirror For Merchants, A Breviate For Bankers, A Treasure For Tradesmen, A Mate For Mechanicks, And A Sure Guide for Purchasers, Sellers, Or Mortgagers of Land, Leases, Annuities, Rents, Pensions, &c. In present Possession or Reversion. And A Constant Concomitant Fitted for All Men's Occasions.' *Octavo*.

London, seventeen-fifty-eight. **Benjamin Donn.** ' A new Introduction to the Mathematicks, being Essays on Vulgar and Decimal Arithmetick. Containing, Not only the practical Rules, but also the Reasons and Demonstrations of them.' *Octavo* (ON).

This good book fulfils to a great extent the profession of the title-page as to demonstration. Donn had a good deal of miscellaneous information. He was, if I remember right, one of Humphrey Davy's early teachers.

London, seventeen-fifty-nine and sixty-four. **Benjamin Martin.** 'A New and Comprehensive System of Mathematical Institutions.' Two volumes. *Octavo in twos.*

Old *Ben Martin* (as his admirers called him) was an able, and in this instance a concise writer. He wrote on every mathematical subject (and never otherwise than well, I believe, except on biography), and a complete set of his works is rarely seen. He was a bookseller.

London, no date, but about seventeen-sixty. **William Weston.** 'Specimens of Abbreviated Numbers.' *Octavo in twos.*

Some supposed new rules for formation and use of decimals.

London, seventeen-sixty. **Jacob Welsh.** 'The Schoolmaster's General Assistant.' In two volumes. *Octavo.*

The author claims a hundred curious discoveries; what they are, I cannot find.

London, seventeen-sixty. **James Dodson.** 'A Plain and Familiar Method for Attaining the Knowledge and Practice of Common Arithmetic.' *Octavo.*

An edition of Wingate's Arithmetic (N) called the nineteenth. But Wingate, who first published it about 1629, would not have known his own book, after the various dressings it received from Kersey, Shelley, and Dodson. One of George Shelley's editions of John Kersey the son's edition of John Kersey the father's edition of Wingate, called the fourteenth, is *London*, seventeen-twenty, *octavo in twos.*

Amsterdam, seventeen-sixty-one. **Isaac Newton.** 'Arithmetica Universalis.' *Quarto.*

This is the edition published by Castiglione, in two volumes, and is the best. The original book, published by William Whiston, consisted of the records laid up in the University Archives of the lectures which Newton delivered as Lucasian professor. [S' Gravesande, in the preface to his edition of *Leyden*, seventeen-thirty-two, says it was published without the author's knowledge, and much to his displeasure: but Whiston, in his Memoirs, says it was with Newton's consent: most likely it was both with his consent, and to his displeasure. Though properly a book on Algebra, and its application to Geometry, yet it does contain a system of Arithmetic. Various alterations, both of matter and arrangement, were made in

H

the (so called) second edition [By Machin, 1722], which are supposed to have been approved, if not furnished, by Newton himself.]
    This information comes from Castiglione. The other editions
I have seen are the original Latin of Whiston, *Cambridge*, seventeen-seven, *octavo*. *London*, seventeen-twenty. ' Universal Arithmetic, ... translated .... by the late Mr. Raphson, and revised
and corrected by Mr. Cunn.' *Octavo in twos.* The same reprinted,
London, seventeen-twenty-eight, *octavo in twos*, called second edition; it is advertised as carefully compared with the correct edition
that was published in seventeen-twenty-two. There is also Wilder's
edition of Raphson's edition, *London*, seventeen-sixty-nine, *octavo*.

    London, seventeen-sixty-five. **J. Randall.** ' An Intro-
duction To so much of the Arts and Sciences, More immediately concerned in an Excellent Education for Trade In its
lower Scenes and more genteel Professions,' &c. *Duodecimo
in threes.*

    Mr. Randall was a quaint man, but his book is well done. It
contains arithmetic, mensuration, and geography ; and ends with a
dialogue between the heavenly bodies, upon their mutual arrange-
ments, in which the earth insists upon being allowed to stand still,
and quotes Scripture like an anti-Copernican, but is brought to
reason by the arguments of the others. This is almost the only
writer I have met with who has given the student a few hints upon
habits of computation. Thus he will not let him say, three and four
are seven, seven and five are twelve, &c.; but only three, seven,
twelve, &c. For, says he, (the example being the addition of some
rents) " as you have this pretty Income, you must talk like a Gentleman to your Figures."

    London, seventeen-sixty-six. **W. Cockin.** ' A Rational
and Practical Treatise of Arithmetic.' *Octavo in twos.*

    Dublin, seventeen-sixty-eight. **Dan. Dowling.** ' Mer-
cantile Arithmetic.' *Octavo in fours.*

    Mostly on exchanges. The author has given the rule for the
instantaneous formation of the fourth and fifth and succeeding
places of the decimal of a pound, which I never saw till now in
any book but my own. The third edition of this book is *Dublin*,
seventeen-ninety-five, *duodecimo in threes*. There is another author
of the same name presently mentioned.

    London, seventeen-seventy-one. **Wm. Rivet.** ' An At-
tempt to illustrate the Usefulness of Decimal Arithmetic.'
Second edition. *Octavo* (small).

    This book contains what I thought no one had given before my-
self, a *complete* head-rule for turning fractions of 1*l.* into decimals
to any number of places. This method, which is much wanted in

commercial arithmetic, is here lost amidst attempts to compute with interminable fractions; things on which real business will never waste a thought.

Amsterdam, seventeen-seventy-three. **Nicolas Barreme.**
' Comptes-faits, ou Tarif général des Monnoies.' *Duodecimo in threes.*

This is a Dutch reprint of the work of a man who has given his name to ordinary mercantile computation in France, even more than Cocker in England. That he was a real person appears from the *privilège* copied into this book, dated Jan. 26, 1760, whereby Louis XV. grants the usual rights over his book to Nicolas Barreme. I have also seen in a catalogue an edition marked seventeen-forty-four. The name became an institution of France, which even the Revolution did not destroy. The Citoyen Blavier published a ' Nouveau Barême,' *Paris*, seventeen-ninety-eight, *octavo*, which he called a *Barême décimal*, in which there is a well-marked distinction between Barême the person and Barême the thing.

London, seventeen-seventy-three. **Thomas Dilworth.**
' Miscellaneous Arithmetic.' In seven parts. *Duodecimo.*

This work is but little known. Its contents are on the calendar; on logarithms; on the rule of three, when the first term is 1*l.* and all the terms are money; on the weather; a collection of riddles, answered, in the midst of which are seriously set forth Bacon's paradoxes on the characteristics of a Christian, and an essay on the education of children. Dilworth had made his name a selling one, and was determined to make use of it.

Shrewsbury, seventeen-seventy-three. **Thomas Sadler.**
' A Complete System of Practical Arithmetic . . . on an entire new plan.' *Duodecimo in threes.*

The newness of plan seems to consist in putting the rules into unintelligible verse, and beating even the older rule-mongers in puzzling plain questions. Thus, a cargo consists of 84, 61, and 35 tons, of which three-fifths is lost; what must each bear of the loss? This is done by first taking $\frac{3}{5}$ of 180, namely 108; then 108 is divided by 180, producing ·6; then each of the parcels is multiplied by ·6.

London, seventeen-seventy-four. **Anth. and Joh. Birks.**
' Arithmetical Collections and Improvements.' Second edition. *Octavo.*

London, seventeen-seventy-four. **Nich. Salomon.** ' The Expeditious Accountant; or, Cyphering rendered *so short*, That Half the Trouble . . . . A VERY CURIOUS WORK, Totally different from all that have preceded it.' *Octavo in twos.*

There is something new, says the author, in almost every rule:

but I cannot find it. The head-rule for the decimals of a pound is introduced.

London, seventeen-seventy-seven. **James Hardy.** 'The Elements or Theory of Arithmetic.' *Duodecimo in threes.*

The author was a teacher at Eton, where, according to common notions, there could have been no such thing at the date above as a teacher of arithmetic. It is true there is an 'ambiguous comma in " Teacher of Mathematics, and writing-master at Eton College." The book is a very creditable one, of great extent, including logarithms, &c.

London, seventeen-eighty. **John Bonnycastle.** 'The Scholar's Guide to Arithmetic.' *Duodecimo.*

The first edition of this well-known work. It had from the beginning algebraical demonstrations attached.

Birmingham, seventeen-eighty-three. **Wm. Taylor.** 'A Complete System of Practical Arithmetic.' *Octavo in twos.*

An enormous book of 600 pages, with arithmetic, mensuration, geography, astronomy, algebra, book-keeping, &c., in the above order.

London, seventeen-eighty-four. **G[eorge] Anderson.** ' The Arenarius of Archimedes . . . from the Greek . . . . the dissertation of Christopher Clavius on the same subject . . .' *Octavo.*

Oxford, eighteen-thirty-seven. **Steph. Pet. Rigaud.** ' On the Arenarius of Archimedes.' *Octavo.*

I choose this edition to introduce the only purely arithmetical work of Archimedes. Its object is to shew that any number, even a universe full of grains of sand, can be easily expressed : a thing by no means likely to be self-evident to a Greek, whose numerical notation was, till Archimedes and Apollonius shewed how it might be extended, far from sufficient for such a purpose. Professor Rigaud, one of the most learned and accurate of our modern inquirers into mathematical history, has given an account of Anderson, and many valuable remarks on his translation, as well as on the subject of it.

London, seventeen-eighty-four. **Thomas Dilworth.** ' The Schoolmaster's Assistant.' *Duodecimo.*

This is the twenty-second edition. By the dates of the commendations prefixed, it would seem that the first edition was published in seventeen-forty-four or forty-five.

Great-Yarmouth, seventeen-eighty-five. **Thos. Sutton.**

'The Measurer's Best Companion ; or, Duodecimals brought to perfection.' *Octavo.*

In truth this is the most elaborate system of duodecimals I have met with.

London, seventeen-eighty-six. **George Atwood.** 'An Essay on the Arithmetic of Factors applied to various Computations which occur in the Practice of Numbers.' *Quarto.*

Edinburgh, seventeen-eighty-six. **John Mair.** 'Arithmetic, Rational and Practical.' *Octavo.* Fourth edition.

The name of Mair, who was rector of the Perth Academy, is highly respected in Scotland. His 'Book-keeping moderniz'd,' of which the *ninth* edition is *Edinburgh,* eighteen-seven, *octavo,* had a great run. Completeness of subjects, and copiousness of examples, characterise both works, which extend to six hundred pages of small print each.

London, seventeen-eighty-seven. **C. G. A. Baselli.** 'An Essay on Mathematical Language ; or, an Introduction to the Mathematical Sciences.' *Octavo in twos.*

It cannot be both, the reader will say : and in truth it is only the second, for arithmetic and algebra, with some good points about it.

London, seventeen-eighty-eight. ' Clavis Campanologia, or a Key to the Art of Ringing.' *Duodecimo in threes.*

As large a list as the present ought to have one book at least on bell-ringing, the whole theory of which is arithmetical. No art has had greater enthusiasts for it. The authors of the present treatise, Wm. Jones, Joh. Reeves, and Thos. Blakemore, are of the number. They record several names of inventors to whom they give words of praise which might apply to Newton or Euler ; among, them is Hardham, who is known to this day by the snuff mixture which he invented and sold in Fleet Street, where his name still remains. I should think that few of those whose noses he has tickled are aware that he may have done the same for their ears.

Oxford, seventeen-eighty-one (one volume) ; *London,* seventeen-eighty-eight (the other). **James Williamson.** 'The Elements of **Euclid,** with dissertations . . . .' *Quarto.*

The arithmetical books of Euclid are here with the rest. I have chosen this edition by which to introduce the name, because it is the only modern translation of *Euclid.* All the works which go by that name are versions thickly scattered with the views of the editors as to what Euclid *ought to have been,* instead of the rendering of *what he was.* For these " many tamperings with his text," a countryman* of Robert Simson has been the first to call them the " per-

---

* Sir William Hamilton of Edinburgh, in his notes to Reid, p. 765.

fidious editors and translators of Euclid ;" a name which, in a sense, they richly deserve. Williamson was a real disciple of Euclid; and he translated so closely, that such words as, not being in the Greek, English idiom renders necessary, are put in Italics. For many editions of Euclid, the reader may consult my article on that name, in Dr. Smith's Biographical Dictionary : and, for the contents of the tenth book, the article *Irrational Quantities* in the *Penny Cyclopædia*.

If the demonstrative system of Euclid had taken as great a hold in arithmetic as in geometry, we should not have had to complain of one of the best exercises of thought being employed for no other purpose than to make machines.

London, seventeen-eighty-eight. **Thomas Keith.** ' The complete practical Arithmetician.' First edition. *Duodecimo in threes.*

London, seventeen-ninety. **Thomas Keith.** ' A Key to the complete practical Arithmetician.' First edition. *Duodecimo in threes.*

London, no date ; about seventeen-ninety (?). **John Duncombe.** ' A new Arithmetical Dictionary.' *Octavo in twos.*

The rules and terms of arithmetic, in alphabetical order.

Calcutta, seventeen - ninety. **Joh. Thos. Hope.** ' A Compendium of Practical Arithmetick.' *Octavo in twos.*

A clear, prolix book, for the Orphan School at Calcutta.

London, seventeen-ninety-one. **William Emerson.** ' Cyclomathesis, or an easy Introduction to the several branches of the Mathematics.' *Octavo.*

These are Emerson's works, collected (not reprinted) in thirteen volumes, with new title-pages. The arithmetic, which is in the first volume, is said by the editor to be of seventeen-sixty-three. Emerson was the writer of many works, which had considerable celebrity : but he was as much overrated as Thomas Simpson was underrated. There is a most amusing life of him prefixed to the collection.

London, seventeen - ninety - one. **Thos. Keith.** ' The New Schoolmaster's Assistant.' *Duodecimo in threes.*

An abridgment of the larger work above mentioned.

Berlin, seventeen-ninety-two. **Leonard Euler.** ' L'Arithmétique raisonnée et démontree, oeuvres posthumes de Léonard Euler, traduite en François par Bernoulli, Directeur de l'Observatoire de Berlin, &c. &c.' *Octavo in twos* (ON).

The editor calls this the first work of Euler in his preface, and posthumous in the title-page. It is, I suppose, a translation of the work which is set down in Fuss's list as 'Anleitung zur Arithmetic, 2 Th. Petersb. 1738. 8,' the *third* of Euler's separate works. It is mostly on commercial arithmetic, and shews that Euler did not, in 1738, consider the old method of division quite exploded.

London, seventeen-ninety-four. **Rich. Carlile.** 'A Collection of one hundred and twenty . . . . Arithmetical, Mathematical, Algebraical and Paradoxical Questions.' *Octavo.*

What *Paradox* is, as a science, I do not know : but the other distinctions are well known. All who know much of the country schools remember that *mathematics* meant geometry, as opposed to arithmetic and algebra. And it was right it should have been so : for neither the schoolboy's arithmetic nor his algebra were *disciplines.*

London, seventeen-ninety-four. **Henry Clarke.** 'The Rationale of Circulating Numbers.' A new edition. *Octavo.*

Another tract on repeating decimals, with some additions on other subjects.

London, seventeen-ninety-four. **Thomas Molineux.** 'The Scholar's Question-book, or an introduction to Practical Arithmetic. Part the second. For the use of Macclesfield School.' *Duodecimo.*

An ordinary school-book on fractions and commercial arithmetic. I never saw the first part. The school-seal, which is engraved on the title-page, gives the learner to understand the mode adopted of explaining difficulties : it displays a pedagogue with a birch-rod in his right hand and a book in his left; illustrative of primary and secondary method. The fourth edition of this second part is *London,* eighteen-twenty-two, *duodecimo.* As may be supposed from the date, the little hint does not appear.

Paris, seventeen-ninety-five. **Agricol de Fortia.** 'Traité des Progressions . . . . précédé par un Discours sur la nécessité d'un Nouveau Système de Calcul.' Third edition. *Octavo.*

The new system of calculation is a proposal to annex to addition, multiplication, and involution, the next step, as the author takes it to be, in the chain of operations. But he is wrong in his use of the analogy. It appears that he was the author of several works on arithmetic.

Dublin, seventeen-ninety-six. **John Gough** (edited by his son). 'Practical Arithmetick in four books.' *Duodecimo.*

This work, I am told, had such extensive currency in Ireland (where it was first published in seventeen-fifty-eight) that the name of the author became almost synonymous with arithmetic ; insomuch that when Professor Thomson's Arithmetic was first published in that country, it went by the name of ' Thomson's Gough.' The second edition appears to have been an augmented and octavo work, afterwards reduced again for schools. It is a book of the ordinary character, and abounds in examples for practice. The last edition I have seen is *Dublin*, eighteen-thirty-one, *duodecimo in threes*, edited by M. Trotter.

New York, seventeen-ninety-seven. **William Milns.** ' The American Accountant.' *Octavo in twos.*

The author seems to have been an emigrant from St. Mary Hall, Oxford. His book has the peculiarity of giving in lieu of answers, the remainders to nine of the answers, for guide in the proof by casting out nines.

Paris, an VI. (seventeen-ninety-seven or ninety-eight). **Condillac.** ' La Langue des Calculs.' *Octavo.*

A posthumous work. The views of a clear-headed mathematician and metaphysician upon the foundation of arithmetic and the formation of its language.

London, seventeen-ninety-eight. **Rich. Chappell.** ' The Universal Arithmetic. *Octavo in twos* (small).

This book deserves notice for the author's attempt to introduce the practice of subtracting in division, without writing down the subtrahend. He versifies his tables, *ex. gr.*

> So 5 times 8 were 40 Scots
> Who came from Aberdeen,
> And 5 times 9 were 45,*
> Which gave them all the spleen.

London, seventeen-ninety-eight. **Francis Walkingame.** ' The Tutor's Assistant ; being a Compendium of Arithmetic, and a complete question-book.' *Duodecimo in threes.*

This is the twenty-eighth edition; when the first was published I do not know, any more than what edition is now current. I should be thankful to any one who would tell me who Walkingame was, and when the first edition was published : for this book is by far the most used of all the school-books, and deserves to stand high among them. I have before me John Fraser's ' Walkingame modernized and improved,' *London*, eighteen-thirty-one, called *seventy-first edition ;* John Little's edition, *London*, eighteen-thirty-nine ; William Birkin's edition, *Derby*, eighteen-forty-three, called the *fifty-first,* and bearing proof that at least seven *Birkins* had appeared ; and Samuel Maynard's edition of F. Crosby's edition, *London*, eighteen-forty-four. All these are *duodecimo.* When editors do not agree

* The North Briton, No. 45.

within twenty as to the number of editions of their author which have been published, that author is surely a man of note.

London, seventeen-ninety-eight. **Wm. Playfair.** 'Lineal Arithmetic, applied to shew the Progress of the Commerce and Revenue of England during the present century.' *Octavo in fours.*

Not arithmetic, but plates arranging the several matters in curves, in the manner now much more familiar than it was then.

London, seventeen-ninety-nine. **Charles Vyse.** The Tutor's Guide, being a complete system of Arithmetic.' *Duodecimo in sixes.* (Tenth edition, edited by J. Warburton ; eleventh in eighteen-one.)

In the same place, year, form, and by the same editor, was published a new edition of the Key. It appears that the first edition was reviewed in the *Monthly Review* for 1771, so that it is probably of the year before. Vyse is one of the most celebrated of the illustrious band who used to adorn the shelves of a country schoolmaster at the beginning of this century ; Vyse, Dilworth, Walkingame, Keith, Joyce, Hutton, Bonnycastle (with Cocker for a lost Pleiad). He is also the poet of the lot : and some of his examples have gone through many other books. The following specimen of the muse of arithmetic should be preserved, as the best known in its day, and the most classical of its kind :

> When first the Marriage-Knot was tied
> Between my Wife and me,
> My Age did her's as far exceed
> As three Times three does three ;
> But when ten Years, and Half ten Years,
> We Man and Wife had been,
> Her Age came up as near to mine
> As eight is to sixteen.
> Now, tell me, I pray,
> What were our Ages on the Wedding Day ?

The book (this tenth edition at least) is crowded with examples, which circumstance makes the Key very large. On the execution there is no remark to make. If a new edition were published, some of the examples must be omitted, as rather opposed to modern ideas of decency.

Paris, eighteen-hundred. 'Séances des Ecoles Normales, recueillies par des sténographes, et revues par les Professeurs. Nouvelle édition.' Thirteen volumes. *Octavo.*

When the Normal School was founded at Paris, in 1794, the professors engaged " pledged themselves to the representatives of the people and to each other" neither to read nor to repeat from memory. Their lectures were taken down in short-hand, and these volumes contain some of them. The professors of mathematics were Lagrange and Laplace : and few persons are aware that the mode in which the two first mathematicians in Europe taught the humblest elements of arithmetic and algebra can thus be judged of. The contents are, so far as these subjects are concerned; vol. I. p. 16, programme and Laplace, arithmetic ; 268, Laplace, arithmetic; 381, Laplace, algebra: II. 116, Laplace, algebra; 302, do. do,: III. 24, Laplace, algebra; 227, Lagrange, arithmetic; 276, Lagrange, algebra; 463, do. do. : IV. 41, Laplace, geometry; 223, Laplace, algebraic geometry ; 401, Lagrange, algebraic geometry : V. 201, Laplace, new system of measures: VI. 32, Laplace, probability : VII. 1, Biot, account of the ' Mécanique Céleste.'

The last three volumes contain the *debates*, or conferences, between the teachers and their pupils, of which there are three in the first of them, on arithmetic and algebra, not at all worth reading.

Taunton, (no date, perhaps before eighteen - hundred). **William Wallis.** ' An Essay on Arithmetic . . . Briefly, Shewing, First, The Usefulness ; Secondly, It's extensiveness ; Thirdly, The Methods of it.'

This is a remonstrance by the author, a teacher at Bridgwater in Somersetshire, against the prevailing modes of teaching arithmetic. The following is an extract: " And I have seen a *Fair-Book* (as 'tis call'd) of a young Man's, about 17 *Years* of Age, who had been 6 *Years* at *School,* but never went through that Rule [of three]: In the same Book I found 132 Questions in *Reduction,* in the working of them were 2680 Figures, which might have been better done in 500, so that there were 2180 *superfluous Ones.* In another Rule I saw an Example, in which were 174 Figures, but might have been done in 23; and one of 80 that might have been done in 12 : In general, I have found in the Boys Books, 3, or 4 Times as many Figures as need be. These *Methods* have so far hindred their Advances in Learning, that amongst 30 Scholars, since I came hither, I have not found one that understood a Rule beyond *Division,* tho' some of them were 14 or 15 *Years* of Age, and had been kept at School, ever since they were capable of being taught."

Paris, eighteen-hundred.   **Condorcet.**   ' Moyen d'apprendre à compter surement et avec facilité.'   Second edition. *Octavo* (duodecimo size).

One of the simplest explanations of the most elementary arithmetic which has ever appeared. It was written in the last days of the author, while hiding from the fate which he only finally avoided by suicide : and the last sheet was hardly finished when his retreat was discovered.

Madrid, eighteen-one.   ' Aritmetica y Geometría práctica de la real Academia de San Fernando.'   *Octavo.*

A very clearly written and printed work.

Buckingham, eighteen-two ; second edition, eighteen-eight. **John King.** ' An Essay intended to establish a new universal System of Arithmetic . . . .' *Octavo.*

The title of the second edition is more modest : it is ' An Essay, or attempt towards establishing . . . . The system is the *octonary* system, in which 10 means eight, 100 is sixty-four, &c.

London, eighteen-three. **Rob. Goodacre.** ' Arithmetic adapted to different classes of learners.' *Duodecimo in sixes.*

The ninth edition of this work, by Samuel Maynard, is *London*, eighteen-thirty-nine, *duodecimo.*

Dublin, eighteen-four. **P. Deighan.** ' A Complete Treatise on Arithmetic, rational and practical.' Two volumes. *Octavo in twos.*

This treatise has a list of a thousand subscribers, and has amused me very much. The old notions of the style of a book were, it seems, not extinct in Ireland a hundred years after they had been exploded in England. The author, who handles his subject ably, puts *philomath* after his name, and is perhaps the last of those who rejoiced in a title which, though self-conferred, its owners would not have changed for F.R.S. It is dedicated " to all those who think that a knowledge of accounts is useful to mankind, from the king on the throne to the lowest subject." It has the praises of the author's friends in prose and poetry, duly prefixed. I quote a few lines from one of the poets, desiring the reader to observe where sophs come from, unsought, and whence Irish authors got their stationery.

> " How many sophs, to sense and science blind,
> Range through the realms of nonsense unconfin'd,
> Unaw'd by shame, and unrestrain'd by law,
> Their labour chaff, and their reward a straw;
> Neglected and despis'd, they sink in shame
> To that *oblivion* whence, *unsought*, they came.
> The muse, indignant, oft with grief has seen
> An author led by ignorance and spleen,
> With snail-paced speed, but unremitting toil,
> In attic chamber waste the midnight oil,
> With waste of paper, loss of ink combined,
> *And pens from public offices purloined.**—
> But DEIGHAN of a more enlightened mind,
> More innate genius, talents more refined," &c.

London, eighteen-five. **Christ. Dubost.** ' Commercial Arithmetic, with an Appendix upon Algebraical Equations.' *Duodecimo in threes.*

Of all works I know professing to be strictly commercial, this has the fullest explanations in words of the rules and processes.

---

* Really I am afraid that there must have been some truth in this. Mr. Thomas Moore gives a translation of the *Pennis non homini datis* of Horace, which shews that he had heard of the thing at least. Of course he can clear himself: at any rate Lalla Rookh has not much the air of having been written with a pen from a public office.

Madras, eighteen-six. **James Brown.** 'A course of Military and Commercial Arithmetic.' *Small octavo size:* no signatures.

As might be expected, this is full on Indian exchanges, weights, and measures.

London, eighteen-six. **William Frend.** 'Tangible Arithmetic; or the Art of Numbering made easy, by means of an Arithmetical Toy which will express every number up to 16,666,665.' *Octavo.* Second edition.

The toy is the Chinese instrument or abacus, called the *Schwan-pan,* for a description of which see Peacock, p. 408.

Paris, eighteen-seven. **J. B. V.** 'L'Arithmétique en-seignée par des moyens clairs et simples.' *Octavo.*

This is in dialogue between a mother and her boy: the author assures us that they are from real life; and it must have been so; for the *ancien officier du génie,* as he calls himself, could no more have written these dialogues than the mother and child could have con-structed a sap. The lady has the awkward name of Madame Épi-nogy; and as the object of the dialogues is to make the child invent, I can find no origin for this name, except the supposition that it is a blundering derivative from ἐπίνοια. Nevertheless the dialogues are exceedingly good.

Paris, eighteen-eight. **F. Peyrard.** 'Oeuvres d'Archi-mède, traduites littéralement, avec un commentaire.' Second edition, two volumes. *Octavo.*

I mention this work, not only for the *Arenarius* already noticed, but for the disquisition by Delambre on the arithmetic of the Greeks, which afterwards appeared in the 'Histoire de l'Astronomie An-cienne,' *Paris,* eighteen-seventeen, *quarto,* two volumes. Delambre was a real reader of the works he cites. He collected his materials from Nicomachus, the *Theologumena,* Barlaam, the two Theons, Ptolemy, Eutocius, Pappus, and Archimedes. I may as well say here what I have to say on those of the above who have not been mentioned elsewhere.

The best edition of Ptolemy is that of Halma, which is a collec-tion of Ptolemy and his commentators, published at different times, and separately, — the whole making distinct works, as well as a set. Of this I know only four volumes, of which the two to the present purpose are *Paris,* eighteen-thirteen and sixteen. 'Κλαυδιου Πτολε-μαιου μαθηματικη Συνταξις . . . .' (Gr. Fr.) *quarto,* two volumes. Brunet says that the commentaries of Theon, and the Κανονες Προχειροι of Ptolemy, are in five more volumes, *Paris,* eighteen-twenty-one, twenty-two, twenty-two, twenty-three, twenty-five, *quarto.*

The commentaries of Eutocius on the works of Archimedes are

to be found in several editions, but best in that of Joseph Torelli, *Oxford*, seventeen-ninety-two, ''Αρχιμηδους τα σωζομενα μετα των Εὐτοκιου 'Ασκαλωνιτου ὑπομνηματων' (Gr. Lat.). *Folio.*

The fragment of the second book of Pappus (the only part of the first two books published as yet, if, indeed, any more exist) is to be found (Gr. Lat.) in the third volume of John Wallis's *Opera Mathematica*, Oxford, sixteen-ninety-nine, *folio in twos.*

London, eighteen-ten. **W. Tate.** 'A System of Commercial Arithmetic.' *Duodecimo.*

A work approximating more nearly to modern business than most of those then in use, in its *additions;* but, like most attempts to improve real commercial arithmetic, wanting the corresponding *omissions.*

Hawick, eighteen-eleven. **Chas. Hutton,** edited by Alex. Ingram. 'A Complete Treatise on Practical Arithmetic and Book-keeping.' *Duodecimo in threes.*

According to Hutton's Catalogue, the fifth edition was in seventeen-seventy-eight, and the twelfth in eighteen-six: and at his death he possessed no edition previous to the fifth. The late Dr. Olinthus Gregory published what he called the eighteenth edition, enlarged, &c. *London*, eighteen-thirty-four, *duodecimo:* and a new edition of Ingram's Hutton, by James Trotter, appeared *Edinburgh*, eighteen-thirty-seven, *duodecimo.*

London, eighteen-twelve. **Thomas Clark.** 'A New System of Arithmetic; including Specimens of a Method by which most Arithmetical Operations may be performed without a Knowledge of the Rule of Three; and followed by Strictures on the Nature of the Elementary Instruction contained in English Treatises on that Science.' *Octavo.*

This is an able attempt to draw public attention to the state of instruction in arithmetic. The author asserts, 1. There is not in the English language, a work of any repute whatever, employed in school education, in which the four fundamental rules of arithmetic are clearly and comprehensively laid down. 2. Not one in which the rules laid down are accompanied by examples so detailed as to remove the difficulties which these rules must present to beginners. 3. None in which the rules and examples for abstract and concrete numbers are kept distinct from each other. 4. There is not a work of this description in which ordinary and decimal fractions are properly arranged. 5. Or in which the rationale of arithmetical operations seems of sufficient importance to the instructor to induce him to. incorporate it with his work. 6. Or in which the principles and algebraical signs used in arithmetic are given and explained at the time when the science requires their introduction.

I

Dublin, eighteen-twelve. **John Walker.** 'The Philosophy of Arithmetic . . . and the Elements of Algebra.' *Octavo*.

Mr. Walker was a good scholar, an excellent mathematician, and a most original thinker. Both this work and that which he published on geometry shew great power.

Sheffield, eighteen-thirteen. **Joseph Youle.** 'The Arithmetical Preceptor . . . . to which is added a Treatise on Magic Squares.' *Duodecimo in threes*.

London, eighteen-thirteen. **Edward Strachey.** 'Biga Ganita; or the Algebra of the Hindus.' *Quarto*.

Bombay, eighteen-sixteen. **John Taylor.** 'Lilawati; or a Treatise on Arithmetic and Geometry by **Bhascara Acharya.**' *Quarto size* (no signatures).

London, eighteen - seventeen. **Henry Thomas Colebrooke.** 'Algebra, with Arithmetic and Mensuration, from the Sanscrit of Brahmegupta and Bháscara.' *Quarto*.

The first work has notes by S. Davis, and is from a Persian version of Bhascara's Sanscrit. The second work is also from the Persian. The third, which contains not only the two works of Bhascara, but also an arithmetical chapter from Brahmegupta, is all from the Sanscrit. It is also pretty copious in selections from the Commentators, and has a large body of dissertation by Colebrooke himself. But it does not entirely supersede the former two, which have likewise valuable annotations.

Edinburgh, eighteen-thirteen. **Elias Johnston.** 'A sure and easy Method of learning to Calculate.' *Duodecimo in sixes*.

This is a translation of the work of Condorcet mentioned under the date 1800.

Paris, eighteen-thirteen. **F. Peyrard.** 'Les Principes fondamentaux de l'Arithmétique.' *Octavo in twos*.

An elegant mixture of arithmetic and algebra, by the editor and translator of Euclid, and the translator of Archimedes.

Lille, eighteen-fourteen. ————. 'Manuel d'Arithmétique ancienne et décimale.' *Duodecimo in threes*.

A small book, in question and answer: a transition book from the old system to the new, containing both, and intended for commercial purposes. At the end are some forms for letters of ceremony and business, and for petitions : and it seems rather strange to English eyes to see that a petition for a son condemned to death for homicide, ranks among the matters which are considered near enough to the ordinary course of business to find a place; and a place which, when opened, gives the option of reading the way of turning francs into roubles.

London, eighteen-fourteen. **S. F. Lacroix.** 'Traité
Élémentaire d'Arithmétique à l'usage de l'école centrale des
quatre-nations.' Tenth edition. *Octavo.*

A well-known work, by one of the most systematic and most
widely circulated of elementary writers. The sixteenth edition was
Paris, eighteen-twenty-three, *octavo*. There was an English translation, *London*, eighteen-twenty-three, *octavo*, anonymous,—an attempt to introduce demonstrative arithmetic into our schools.

The third edition of an American translation, by John Farrar,
appeared *Cambridge* (*U. S.*), eighteen-twenty-five, *octavo in twos.*

Dublin, eighteen-fourteen. **R. F. Purdon.** 'Theory of
some of the Elementary Operations in Arithmetic and Algebra.'
Octavo.

Oxford, eighteen-fourteen. **Charles Butler.** 'An Easy
Introduction to the Mathematics.' *Octavo.*

This book fulfils the promise of the title-page well, and has been
frequently cited for the historical introductions to the several subjects, which are very good, and, as parts of a learner's course, un-
exampled.

London, eighteen-fifteen. **J. Carver.** 'The Master's
and Pupil's Assistant.' *Duodecimo in threes.*

The author of this work, dependent as the sale of it was on
teachers, has had the sense and courage to say, that questions with
answers are *for the benefit of the masters* and the *injury of the pupils.*
It is dated from *Belgrave House, Pimlico,*—a name and site which
might puzzle an antiquary a century hence.

Paris, eighteen-sixteen. **Bezout.** 'Traité d'Arithmétique.'
Eighth edition. *Octavo.*

A work of a somewhat older stamp than those of Lacroix and
Bourdon. The eleventh edition was *Paris*, eighteen-twenty-three,
octavo, edited by A. A. L. Reynaud, with notes and a table of
logarithms; the notes a separate work, with another title-page.
The next year appeared the twelfth edition of Reynaud's own work,
Paris, eighteen-twenty-four, 'Traité d'Arithmétique,' *octavo*, also
augmented by a table of logarithms. This work enters rather more
on the theory of numbers.

London, eighteen-sixteen. **Thos. Taylor.** 'Theoretic
Arithmetic, in three books; containing the substance of all
that has been written on this subject by Theo of Smyrna, Nicomachus, Jamblichus, and Boetius. Together with some remarkable particulars respecting perfect, amicable, and other
numbers, which are not to be found in the writings of any

ancient or modern mathematicians. Likewise, a specimen
of the manner in which the Pythagoreans philosophised about
numbers ; and a development of their mystical and theological
arithmetic.' *Octavo.*

Edinburgh, circa eighteen-sixteen. **A. Melrose** (edited
by A. Ingram). 'A Concise System of Practical Arithmetic.
Second edition. *Duodecimo in threes.*

Horner refers to this book as containing what is nearly an anti-
cipation of his method, in the case of the simple cube root.

London, eighteen-seventeen. **Thos. Preston.** 'A New
System of Commercial Arithmetic . . . . a perfect, a permanent
and universal ready reckoner.' *Duodecimo in threes.*

Perhaps the plan of this book is partly taken from one published
by Girtanner in seventeen-ninety-four (*Penny Cycl. Suppl.* Tables),
in which the logarithms of numbers and certain intermediate fractions are given. But, in the main, it is an application of the same
principle as that which has long been used in astronomical logarithms, namely, giving the logarithms of integers, with a column in
which those integers, considered as seconds, are turned into degrees,
minutes, and seconds. In this way are given the logarithms of pence
up to 130*l.* ; of pounds up to 7 cwt. 16lbs. ; of twelfths up to 333⅓ ;
of sixteenths up to 312½ ; of sixteenths of gallons, considered as
fractions of a tun of 236 gallons, for seed-oils ; the same for a tun of
256 gallons, for fish-oils and wines ; of pounds of 120 to the cwt. up
to five tons ; of grains, up to 25 oz. troy ; and two for days and for
pounds at 5 per cent, by which one operation gives the interest for
days on any number of pounds up to 2600*l.*

Leipsic, eighteen-seventeen. **Fred. Astius.** 'Theologumena Arithmeticæ . . . . Accedit **Nicomachi** Geraseni Institutio Arithmetica.' *Octavo.* (See p. 17.)

These θεολογούμενα have been attributed to Jamblichus and to
Nicomachus : but they seem rather to consist of extracts from
Nicomachus, Anatolius, and others. They are explanations of the
Pythagorean and Platonic opinions on numbers ; and form a very
good accompaniment for the works of Nicomachus. The notes are
full and good.

London, eighteen-eighteen. **George G. Carey.** A
Complete System of Theoretical and Mercantile Arithmetic.'
Octavo.

A commercial book with a table of logarithms in it is rare in the
nineteenth century.

Edinburgh, eighteen-eighteen. **William Ritchie.** 'A
System of Arithmetic . . . . and a Course of Book-keeping.
Duodecimo in sixes.

A book of much greater merit than could be guessed from its pretensions or its notoriety : and, for its size, one of the most comprehensive I have met with. Its author was a little (about twenty years) in advance of his age, and the greater part of the edition was sold as waste paper.

London, eighteen-nineteen. 'Philosophical Transactions.' *Quarto*. **W. G. Horner.** 'A New Method of Solving Numerical Equations of all orders by Continuous Approximation.'

London, eighteen-thirty. **Thos. Leybourn.** 'New Series of the Mathematical Repository.' Volume five. *Octavo in twos.* **W. G. Horner.** 'Horæ Arithmeticæ.'

The first-mentioned paper contains the most remarkable addition made to arithmetic in modern times, the value of which is gradually becoming known. On this subject I may refer to Mr. Horner's paper on Algebraic Transformation in the Mathematician (vols. i. and ii. various numbers); to J. R. Young, 'An elementary treatise on Algebra,' *London*, eighteen-twenty-six, *octavo*, as the first elementary writer who saw the value of Horner's method ; J. R. Young, 'Theory and solution of Algebraical Equations of the higher orders,' *London*, eighteen-forty-three, *octavo ;* Thos. Stephens Davies, 'A Course of Mathematics . . . . by Charles Hutton,' twelfth edition, *London*, eighteen-forty-three, *octavo ;* T. S. Davies, 'Solutions of the principal questions in Dr. Hutton's Course of Mathematics,' *London*, eighteen-forty, *octavo ;* Peter Gray, four papers in the Mechanic's Magazine for March, eighteen-forty-four ; A. De Morgan, 'Notices of the progress of the problem of Evolution,' in the *Companion to the Almanac* for eighteen-thirty-nine, with the two articles headed 'Involution and Evolution' in the Penny Cyclopædia, and in the Supplement (eighteen-thirty-eight and forty-five), and a Letter to the Editor of the Mechanic's Magazine, published in that work for February, eighteen-forty-six.

In connexion with this subject I ought to mention Mr. Thomas Weddle's 'New simple and general method of solving numerical equations of all orders,' *London*, eighteen-forty-two, *quarto*. This is an organised process, in which the principle of each step is the correction of the preceding result by multiplication, not by addition.

Edinburgh, eighteen-twenty. **John Leslie.** 'The Philosophy of Arithmetic.' *Octavo*.

I have spoken of this work in the Introduction. P. 373, 405, 411, 477.

Vienna, eighteen-twenty-one. **Geo. Fred. Vega.** 'Vorlesungen über die Rechenkunst und Algebra.' *Octavo*.

The fourth edition (with preface dated seventeen-eighty-two, which I presume to be the date of the first) of the Arithmetic of the celebrated editor of the greatest modern table of logarithms.

Leeds, eighteen-twenty-three. **[Walker ?]** 'Elements of Arithmetic for the use of the Grammar School . . . .' *Duodecimo in threes.*

An excellent little work, which I suppose I am right in attributing to Mr. Walker, and calling it the first edition of the work presently mentioned.

London, eighteen-twenty-three. **Thomas Taylor.** 'The Elements of a New Arithmetical Notation, and of a New Arithmetic of Infinites.' *Octavo in twos.*

A curious attempt at establishing a theory of infinites, the unit of which is $1+1+1+$ &c. ad inf. Those who know nothing of Taylor *the Platonist*, should read his life in the *Penny Cyclopædia*. To re-establish Plato and Aristotle (in some sense even their very mythology) was the uniform endeavour of a long life and a most voluminous course of authorship. P. 424.

London, eighteen-twenty-four. **Jas. Darnell.** 'Essentials of Arithmetic, or Universal Chain.' *Duodecimo.*

This treatise reduces most questions to the form known as the *chain-rule.* Several writers have since advocated this plan.

London, eighteen-twenty-five. **J. Joyce.** 'A System of Practical Arithmetic.' *Duodecimo.*

A well-known book. The preface is dated eighteen-sixteen, which I suppose to be the date of the first edition.

London, eighteen-twenty-six. **J. R. Young.** 'An Elementary Treatise on Algebra.' *Octavo.*

I enter this work here as that of the first elementary writer who saw the value of Horner's (or, as he then called it, Holdred's) method of solving equations. This is, I believe, the first of the series of widely-known and much-used works by the same author.

London, eighteen-twenty-six. **Chas. Pritchard.** 'Illustrations of Theoretical Arithmetic.' *Duodecimo in sixes.*

A demonstrated system of Arithmetic, at a time when there were few such things in English.

London, eighteen-twenty-seven. **H.** and **J. Grey.** 'Practical Arithmetic.' Eighth edition. *Duodecimo.*

A book of concise rules for special cases.

London, eighteen-twenty-seven. **George Walker** (Master of the Grammar School at Leeds). 'Elements of Arithmetic, theoretical and practical, for the use of the Grammar School, Leeds.' Third edition. *Duodecimo.*

A clear and excellent work, written by a man of real science. I doubt whether the peculiarity of the work, the introduction of the distinction of integers and decimal fractions at the very outset, be judicious : but it has had advocates of powerful name.

Paris, eighteen - twenty - eight. **Bourdon.** ' Elémens d'Arithmétique.' Sixth edition. *Octavo.*

More complete than Lacroix in details. Bourdon was an excellent elementary writer, both on arithmetic and on algebra. I began my career as an author, by a translation of part of his work on algebra.

London, eighteen - twenty - eight. **John Bonnycastle.** ' The Scholar's Guide to Arithmetic . . . . enlarged and improved by the Rev. E. C. Tyson.' *Duodecimo.*

What edition this is I do not know.

London, eighteen-twenty-eight. 'An Abridgment of the Arithmetical Grammar . . . . By Catechetical Scrutiny.'

I have never met with the larger work.

Newcastle, eighteen - twenty - eight. **William Tinwell** (edited by James Charlton). ' Treatise on Practical Arithmetic, with Book-keeping, by single and double entry.' Twelfth edition. *Duodecimo in threes.*

Dublin, eighteen - twenty - eight. **John Garrett.** 'An Essay on Proportion.' *Octavo.*

Mostly arithmetical, with something on the connexion of number and magnitude.

London, —— **George Peacock.** ' Arithmetic.'

This is from the *Encyclopædia Metropolitana.* As part of the first volume of the *Pure Sciences,* the date is eighteen-twenty-nine : but it was separately published, in the parts, in eighteen-twenty-five or twenty-six. This is the article mentioned in the Introduction. I subjoin a few remarks, either in correction of some slips of the pen, or in addition to what has been said. The paging is that of the *Encyclopædia Metropolitana.*

P. 402, note. A sexagesimal table is sometimes pasted into a work to which it does not belong, by some old owner.

P. 404. For ' Regiomontanus, in his Opus *Palatinum* de Triangulis,' read ' in his work *de Triangulis ;*' and for ' as we learn from the relation of Valentine Otho, *in his preface to that work,*' read ' in his preface to the *Opus Palatinum.*' But further observe, that though Regiomontanus did change the usual sexagesimal radius into a decimal one, no decimal tables of his were *published* until some time after Apian had published decimal tables of his own. At least, after much research, I can find none even mentioned. It is a mistake (a

universal one) to say that decimal sines were published with the
work *de Triangulis*, in 1533.   It arises from such tables being in the
second edition of that work, in 1561.

P 411.   Add that Chaucer, in his work on the *Astrolabe*, makes
augrime figures to be exclusively the Arabic numerals.

P. 414. The work of *Pacioli* (not Paccioli, though very often so
spelt) was published in 1494 (not 1484).   Canacci, not Caracci.

P. 419 (note).   Besides the *Fasciculus Temporum* (of which, by
the way, there are editions dated 1474, and two or three without
date, probably earlier, according to Hain), there are the *Almanac*
and *Éphemerides* of Regiomontanus, described in the *Companion
to the Almanac* for 1846, with extensive masses of Arabic numerals.

P. 425.   Note that there is an express commentary by Jam-
lichus on Nicomachus, and that the former probably did not write
the *Theologumena* at all.

P. 434.   I have no doubt that Dr. Peacock has authority for
saying that the old English Arithmeticians called the then common
mode of dividing the *scratch* way (as indeed it was); but I have
never met with the phrase.

P. 438 (note).   It was quite common, even before the invention
of printing, to speak of *Algus* as the inventor of decimal notation, or
Algorithm : and several of the ornamented title-pages of Simon de
Colines, the successor of Henry Stephens, have figures of Ptolemy,
Orpheus, Euclid, and *Algus*, on one side, opposite to the Muses of
Astronomy, Music, Geometry, and *Arithmetic*, on the other : as if
paired for a dance.

P. 440. See what I have said on the *Disme* of Stevinus (1585);
and note that Simon of Bruges is Stevinus himself.

P. 442.   There is nothing about decimal fractions in Hunt's
Handmaid.

P. 444.   I suggest as the derivation of *furlong*, a corruption of
forty-long.   The ordinary derivation, *furrow-long*, can hardly have
a foundation in fact; for the length of a furrow depends upon that
of a field.

P. 454.   For ' Wingrave' read ' Wingate.'   As to the note, I
differ greatly (as appears elsewhere) from Dr. Peacock's opinion of
Cocker.   With respect to the last sentence : ' It may be worth while
observing that this modest and useful book is not honoured with
poetical recommendations ;' it would seem that the copy consulted
wanted the frontispiece, on which, under the portrait of Cocker, are
these lines :

> " Ingenious COCKER ! (now to Rest thou'rt Gone)
> Noe Art can Show thee fully but thine own,
> Thy rare *Arithmetick* alone can show
> Th' vast *Sums* of Thanks wee for thy Laboure owe "

P. 464.   For ' James Peele' read ' John Mellis,' and alter the
date to 1588, as above.   See on Stevinus what I have said above.

P. 471.   The *leuca* was originally a measure of the Gauls; and
it may be much doubted whether the measure was ever out of
France since the time of Cæsar.   See the article *League* in the
Penny Cyclopædia.   Surely Dr. Peacock must have written 'France'

meaning to have written 'England.' This alteration of one word
makes every thing right.

The small number of inaccuracies noted above are all I have
been able to lay my hands on, in going through Dr. Peacock's most
valuable article, with the materials for this list about me. I may add,
that in two, or perhaps three instances, I remember to have found
an edition mentioned as the earliest which is not so. But to this
every writer is subject who has the courage to attempt history upon
such bibliographical guides to the sources as now exist.

London, [eighteen-twenty-nine]. **James Parker.** 'Arithmetic and Algebra.' *Octavo*.

An early work of the Society for the Diffusion of Useful Knowledge. Arithmetical and algebraical demonstrations are connected.

London, eighteen-thirty. **Joh. White.** 'An Elucidation
of the Tutor's Expeditious Assistant.' *Duodecimo*.

The author claims as a new discovery in arithmetic, the notion
of forming examples in which the figures of the answer have some
perceptible relation, unknown to the pupil, but known to the teacher,
who can thereby see whether it is right or wrong. All the questions set, for instance, in multiplication have digits which descend
by twos, as in 86420, 20864, 75319, &c. So that it is gravely proposed that the pupil shall be trained upon selected questions in
which the answers have all the figures odd, or all even. A pupil so
trained would be in danger of being led to think that 3728 could
not be the product of any numbers.

Edinburgh, eighteen-thirty. **Adam Anderson.** 'Arithmetic.' *Quarto*.

This is the article so headed in Brewster's Cyclopædia, vol. ii.
part 1. It has a fair amount of arithmetical history, though hardly
enough for the extent of the work of which it forms a part.

London, eighteen-thirty-one. **Frederic Rosen.** 'The
Algebra of **Mohammed Ben Musa.**' *Octavo in twos*. English and Persian.

This algebra is very arithmetical. It was published by the
Oriental Translation Fund. M. Libri, in the work presently mentioned (vol. i. pp. 253-297), has given a long extract from old Latin
translations now in manuscript in the Royal Library at Paris. This
book passes for the one by means of which Leonard of Pisa was the
first European who learnt algebra, or at least who wrote upon it.

Derby, eighteen-thirty-three. **Samuel Young.** 'A
System of Practical Arithmetic.' *Duodecimo*.

A work with great force of rules, and examples in machinery,
manufactures, &c.

Springfield (*U. S.*), eighteen-thirty-three. **Zerah Colburn.** 'A Memoir of Zerah Colburn, written by himself . . . with his peculiar methods of calculation.' *Duodecimo in threes.*

A great many will remember that in 1812 and 1815 two young boys, Zerah Colburn and George Bidder, astonished every one by a power of rapid mental calculation to which the most practised arithmeticians could not make the least approach. Mr. Colburn was, in 1833, a minister among the Methodists in the United States; Mr. Bidder, there is little occasion to say, is a civil engineer in England. The peculiarity of Colburn was, that he could extract roots and find the factors of numbers, to an extent which the mathematician himself had no organised rule for doing. Speaking in the third person, Mr. Colburn says: " Some time in 1818, Zerah was invited to a certain place, where he found a number of persons questioning the Devonshire boy (Mr. Bidder). He displayed great strength and power of mind in the higher branches of arithmetic ; he could answer some questions that the American would not like to undertake ; but he was unable to extract the roots and find the factors of numbers." This treatise contains an account of Colburn's methods.

London, eighteen-thirty-seven. **Daniel Harrison.** 'A New System of Mental Arithmetic, by the acquirement of which all numerical questions may be promptly answered without recourse to pen or pencil . . . . an entirely novel method of reducing the largest sums of money . . . . to their lowest denominations by means of original quadrantal rationale' . . . . *Duodecimo in threes.* Second edition.

The pretensions of this work are manifestly exaggerated ; but the methods are ingenious. Books of mental arithmetic choose their own examples, and thus make their own rules work well. The *quadrantal* method (why so called I cannot guess) is making use of a rule for turning pounds, &c. into farthings, which requires the learning of a table ; and then making such applications as the following :—A pound troy contains six times as many grains as there are farthings in the pound ; therefore the grains in 4 pounds troy are the farthings in six times as many pounds sterling. Some of these rules would really become very effective, if the method of decimalising the parts of a pound were used. Multiply the pounds by a thousand, and subtract 4 per cent, and the result is the number of farthings. Thus :

$$£1763 \; . \; 17 \; . \; 9\tfrac{3}{4} \text{ is } 1763\cdot89062$$

	1763890·62	
4 per cent	70555·62	
Subtract	1693335	= No. of farthings in £1763 . 17 . 9¾

Berwick-upon-Tweed, eighteen-thirty-eight. **James Gray** (edited by Wm. Rutherford). 'An Introduction to Arithmetic.' *Duodecimo in threes.*

Mr. Rutherford says this neat little work went through more than forty editions in the half century preceding this publication. How many works on arithmetic there must be which I have never seen! I never met with any one of the forty.

Paris, eighteen-thirty-eight, thirty-eight, forty, forty-one. **Guillaume Libri.** 'Histoire des Sciences Mathématiques en Italie, depuis la renaissance des lettres jusqu'à la fin du xvii⁰ siècle.' *Octavo.*

The first volume of this work was printed *Paris,* eighteen-thirty-five, *octavo,* but it was never sold; the whole impression (except a few copies distributed as presents) was burnt. The work is yet un-finished; four volumes, dated as above, being all that have appeared. It is a valuable history as concerns arithmetic, both in text and notes; the latter contain—the *liber augmenti et diminutionis* compiled by Abraham (supposed to be Aben Ezra) *secundum librum qui Indorum dictus est* (vol. i. pp. 304-376)—Account of an extract from the *Liber Abbaci* of Leonard of Pisa, written in twelve-hundred-and-two (vol. ii. pp. 287-304) — the *practica Geometriæ* of the same author, being the whole of his algebra, written in twelve-hundred-and-twenty (vol. ii. pp. 305-476) — extracts from Pacioli (vol. iii. pp. 277-294) : besides many other matters not relating to arithmetic. As this work stands, it can be little used for want of an index.

London, eighteen-thirty-eight. **Thos. Keith.** 'The Complete Practical Arithmetician.' *Duodecimo.*

The editor (Mr. Maynard, the mathematical bookseller, who ought to know) calls this the 12th edition : it has a deservedly high character among books of rules only, for precision and completeness. Keith says that the bare names of those who have written on arithmetic in England, from the time of Wingate, would fill a moderate volume. I suspect not : after having examined every source within my reach, and got only 1500 names, out of all times and countries, I should think it impossible that Keith could have known of 300 Englishmen, within the limits mentioned.

London, eighteen-forty. **Thomas Stephens Davies.** ' Solutions of the Principal Questions of Dr. Hutton's Course of Mathematics ' *Octavo.*

This work is also a running comment on the original, and it has a specific right to be in this list from its abounding with instances of Mr. Horner's various methods of manipulating arithmetical and more particularly algebraic questions, independently of his celebrated method of solving equations. In eighteen-forty-one the same author published an edition of Hutton's Course itself.

London, eighteen-forty-two. ' Encyclopædia Britannica.' Vol. III. *Quarto.*

The article on arithmetic was written, I suppose, by Leslie ; or if not, by a follower of his views. It is meagre in its history.

London, eighteen-forty-three. **Alfr. Crowquill** (as he calls himself). ' The Tutor's Assistant, or Comic Figures of Arithmetic.' *Duodecimo in sixes.*

Walkingame's arithmetic, with comic woodcuts : but the subject will not furnish good materials. Under ' Subtraction of time,' for instance, is represented the stealing of a watch ; as a heading for *Troy weight,* the *wooden horse ;* and for *measure of capacity,* a phrenologised head. The worst of it is, that the joke will remain on hand too long for the learner : a picture of a gamekeeper producing a hare from a poacher's pocket must stare him in the face, as an illustration of ' proof,' through twenty-two mortal questions of addition. A comic arithmetic, with the cuts in illustration of the examples for exercise, would give the artist much fairer play.

Liège, eighteen-forty-four. **H. Forir.** ' Essai d'un cours de Mathématiques à l'usage des élèves du collége communal de Liége . . . . . Arithmétique.' Eighth edition. *Octavo in fours.*   .

A good and well-printed treatise, with many examples.

London, eighteen-forty-five. **Thomas Tate.** 'A Treatise on the First Principles of Arithmetic.' *Duodecimo.*

Two pages at the end, on the use of the properties of nine in constructing questions for pupils, are well worthy the attention of teachers.

London, eighteen-forty-six. **A. De Morgan.** ' The Elements of Arithmetic.' Fifth edition, with Appendixes. *Duodecimo in threes.*

The previous editions were eighteen-thirty, thirty-two, and thirty-five, *duodecimo,* and eighteen-forty *in threes.*
Books of bibliography last longer than elementary works ; so that I have a chance of standing in a list to be made two centuries hence, which the book itself would certainly not procure me.

The following are some additional works more briefly described.   Here $2+2$, $4+4$, $6+6$, mean quarto in ones, octavo in twos, duodecimo in threes.

Date.	Place.	Author.	Leading Word of Title.	Form.
SEVENTEEN-				
-thirty-four .	London	ALEX. WRIGHT . .	Fractions	6+6.
-fifty-four . .	Exon.	JOSEPH THORPE . .	Treatise	4+4.
-fifty-seven .	London	R. GADESBY . . .	Decimal	4+4.
-sixty-one . .	London	JOHN DEAN. . . .	Rule of Practice	8 (2d ed.)
-sixty-two . .	London	RICH. RAMSBOTTOM .	Fractions anatomised . . .	6+6.
-sixty-six . .	Sheffield	JOHN EADON . . .	Guide	6+6.
-sixty-six . .	Birmingham . . .	WM. CRUMPTON . .	Decimals and Mensuration .	6+6.
-seventy . .	Exeter	G. DYER	School-Assistant	12.
-seventy-one .	London	WM. SCOTT	New System	8.
-eighty-eight	Birmingham . . .	WM. TAYLOR . . .	Guide	6+6.
-eighty-nine .	Edinburgh . . .	WM. GORDON . . .	Institutes	4+4.
EIGHTEEN-				
-four . . .	Bingham	W. BUTTERMAN . .	Dialogue	6+6.
-five . . .	Shrewsbury . . .	GEO. BAGLEY . . .	Assistant	2+2.
-eight . . .	Paris	EDM. DEGRANGE . .	Pratique	8.
-eleven . .	London	JOH. HARRIS WICKS	Merchant's Companion . .	6+6.
-eleven . .	Birmingham . . .	J. RICHARDS . . .	Practical	6+6.
-twelve . .	Paris	J.CL.OUVRIERDELILE	Appliquée au Commerce . .	8.
-sixteen . .	Edinburgh . . .	R. HAY	Beauties	8.
-seventeen .	London	JAMES MORRISON .	Concise System	12.
-eighteen . .	London	JOH. MATHESON . .	Theory and Practice . . .	12.
-eighteen . .	Dundee	JAMES NICOLSON . .	Modern Accountant . . .	6+6.
-twenty . .	Glasgow	C. MORRISON . . .	Young Lady's Guide . . .	6+6.
-twenty . .	Paris	J. B. JUVIGNY . . .	Application au Commerce .	8.
-twenty-two .	Bedford	W. H. WHITE . .	Complete Course	6+6.
-twenty-three	Canterbury . .	{ H. MARLEEN and W. SMEETH . . }	Assistant	6+6.
-twenty-three	Lausanne	EM. DEVELEY . . .	D'Emile	8.
-twenty-four	Cork	JOH. PENROSE . . .	Treatise	6+6.
-twenty-six .	London	ANTH. PEACOCK . .	Mental	12.
-twenty-six .	London	J. W. HOAR. . . .	Examinations	18.
-twenty-seven	London	WM. PHILLIPS . .	New and concise System .	12.
-twenty-eight	London	S. P. REYNOLDS . .	Practical	4+4 (small).
-twenty-eight	London	ROB. FRAITER . . .	Short System	6+6 (2d ed.)
-twenty-eight	[London]	R. W. WOODWARD .	Guide	12.
-twenty-eight} [circa] }	London	ANONYMOUS . . .	Intellectual (Pestalozzian) .	6+6.
-twenty-nine	Paris	H. L. D. RIVAIL . .	D'après Pestalozzi . . .	12.

Date.	Place.	Author.	Leading Word of Title.	Form.
EIGHTEEN-				
-twenty-nine	Paris	A. SAVARY	Arithmétique	8.
-twenty-nine	London	ANONYMOUS . . .	Concise Arithmetician . .	6+6.
-twenty-nine	London	DANIEL DOWLING .	Improved System	12.
-twenty-nine	London	JAMES PERRY . . .	Middle Stage and Key . .	6+6.
-thirty . . .	London	W. PUTSEY	Practical	6+6.
-thirty-one .	London	GEO. HUTTON . . .	Theory and Practice . . .	12.
-thirty-two .	Edinburgh . . .	ROB. CUNNINGHAM .	Text-Book	6+6.
-thirty-four .	London	RICH. CHAMBERS . .	Introduction	6+6.
-thirty-four .	Newcastle-on-Tyne	THOS. THOMPSON . .	Business Book	4+4.
-thirty-four .	London	CHRIST. KNOWL. SOC.	Taught by Questions . . .	8 (square).
-thirty-six .	London	RICH. MOSLEY . . .	Elements	12.
-thirty-six .	Edinburgh . . .	JOH. DAVIDSON . .	Guide	6+6.
-thirty-seven	Boston (U. S.) . .	JOS. BARTRUM . . .	Arithmetic	6+6.
-thirty-nine .	London	J. FELTON	Improved Method . . .	6+6.
-forty . . .	Cambridge . . .	W. C. HOTSON . . .	Principles	6+6 (2d ed.)
-forty-three .	London	MERCATOR	Expeditious Calculation . .	4+4.
-forty-five .	London	W. WATSON	School	6+6(4th ed.)
Without date	London	ADAM TAYLOR . .	[Useful Arithmetic] & Sequel	12.
————	London	HENRY AULT . . .	Requisites of Business . .	6+6 (pt. 1).
————	Dartmouth . . .	WM. JARVIS . . .	Catechism	8 (small).

ADDITIONS.

———

No place, date, nor author's nor printer's name. 'Algorithmus Integrorum Cum Probis annexis.' *Quarto in threes.*

I think this is the oldest book in my list. The type is of that manuscript appearance which is so common in books printed before 1480 : and this one can hardly be later than 1475, and may be ten years earlier. The matter is that of a writer prior to the school of Borgi and Pacioli. Though the nine digits are given, they are not used. The rules only are given, and they are Sacrobosco's set; for which see p. 15.

No place nor date. 'Algorismus novus de integris compendiose sine figurarum (more Italorum) deletione compilatus. artem numerandi omnemque viam calculandi enucleatim brevissimè edocens. una cum algorismis de minuciis vulgaribus videlicet et phisicalibus. Addita regula proportionum tam de integris quam fractis que vulgo mercatorum regula dicitur. Quibus habitis quivis modica adhibita diligentia omnem calculandi modum facillime adipisci potest.' *Quarto in fours.*

Three different editions of this work are mentioned by bibliographers, all without dates. Numeration is called *prima species;* addition, *secunda species;* and so on. There are slight examples, except upon the square root, in which only results are given. It is something between such a work as the last, with rules only, and that of Borgi, to which I now come, in which there is decided force of examples.

Venice, fourteen-eighty-eight. **Piero Borgi.** Work on Arithmetic without title, as follows. *Quarto in fours.*

This is one of the editions alluded to in page 2, and for the power of mentioning it from inspection (as well as the works of Ghaligai and Texeda following) I am indebted to Dr. Peacock. It begins with the following verses :

> Chi de arte matematiche ha piacere
> Che tengon di certeza el primo grado
> Avanti che di quelle tenti el vado
> Vogli la presente opera vedere
> Per questa lui potra certo sapere
> Se error sara nel calculo notado

Per questa esser potra certificado
A formar conti di tutto [*sic*] maniere
A merchadanti molta utilitade
Fara la presente opera e afatori
Dara in far conti gran facilitade
Per questa vederan tutti li errori
Ede iquaterni soi la veritade
Danari acquisterano e grandi honori
           In la patria e de fuori
Sapran far le rason de tutte gente
Per le figure che son qui depente.

After a page of description of symbols, comes 'Qui comenza la nobel opera de arithmeticha ne laqual se tracta tute cosse amercantia pertinente facta et compilata per Piero borgi da Veniesia.' At the end is ' Stampito in Veniexia per zovanne de Hallis 1488.'

The author begins by saying that there are plenty of sufficiently good masters, and not less abundance of most excellent authors. His arrangement of the five *acts* of arithmetic is strange; they are numeration, multiplication, partition, summation, and subtraction. He uses the word *million*, and his numeration goes to millions of millions of millions. He then proceeds to multiplication, and forms a table of the nine multiples of 1, 2, ... 10, with those of 12, 16, 20, 24, 32, and 36. He then points out how to prove processes by casting out sevens and nines; in doing which he shews how to divide by 7, and uses the sum of the digits to find the remainder to 9. Then follow multiplications done each in one line, even when the multiplier has three or four figures. After this follows the common method, in which the final process is addition, done without any formal rule (in fact, addition is not made a formal rule any where). Division is said, at the head of a chapter, to be done in three ways; but only two are given: that *per cholona*, where the whole is done in one line, and that *per batelo*, which is what I have always designated by (O). Subtraction is then explained; and after some applications, the rules for fractions (which are expressed in the same way as now) are introduced. The rule of three (*riegola del tre*) is given, and a large number of money applications.

Paris, fourteen-ninety-six. ' In hoc opere contenta. Arithmetica decem libris demonstrata Musica libris demonstrata quatuor Epitome in libros arithmeticos divi Severini Boetii. Rithmimachie ludus qui et pugna numerorum dicitur.' *Folio in fours.*

This is the first edition of the work mentioned in page 10: so that I was wrong in page 4. I was misled by bibliographers mentioning this as an edition of Jordanus only. The copy before me comes to me in two different parts, though the register at the end (which has the first words of all the sheets) proves that these parts belong to one work. It would be odd if it should have been customary to divide it in this manner, so as to lead to the above imperfect description. The corrector of the press was David Lauxius, and the printers John Higman and Wolfgang Hopilius.

Venice, no date, but apparently before fifteen-hundred. 'Dabaco che insegna a fare ogni ragione mercadantile: et a pertegare le terre con larte di la geometria: et altre nobilissime ragione straordinarie con la Tariffa come respondeno li pesi et monede de molte terre del mondo con la inclita citta de Vinegia. El qual Libbro se chiama Thesauro universale.' *Octavo.*

The order of the operations is as in P. Borgi. The book is very full of florid ornaments and woodcuts.

Among books of the fifteenth century which I have had no opportunity of seeing are the following, described by Hain and others:

Without date. 'Arithmeticæ Textus communis, qui pro Magisterio fere cunctis in gymnasiis ordinarie solet legi, correctus corroboratusque perlucida quadam ac prius non habita commentatione a Conrado Norico. Lipsiæ per Martinum Herbipolensem.' *Folio.*

Without date, *quarto,* with the mark of Martin of Wurtzburg (Herbipolis). 'Algorithmus linealis,' beginning 'Ad evitandum multiplices Mercatorum errores &c.'

Treviso, fourteen-seventy-eight, *quarto.* 'L'arte del Abbacho, Practica molto bona et utile a chiachaduno qui vuole uxare l'arte merchandantia.'

Watt mentions an edition, *Rome,* fourteen-eighty-two, *quarto,* of the 'Ludus Arithmomachiæ,' attributed to John (he should have said William) Shirewood, bishop of Durham. Hain does not mention this.

Babenberg, fourteen-eighty-three, *duodecimo.* 'Rechnungsbüchlein.'

Cologne, fourteen-eighty-five, *quarto.* 'Ars numerandi, seu compendiosus tractatus de dictionibus numerabilibus.'

Strasburg, fourteen-eighty-eight, *quarto.* Anianus. 'Compotus manualis magistri aniani. metricus cum commento Et algorismus.

Leipzig, fourteen-eighty-nine, *octavo.* 'Rechnung auf alle Kaufmannschaft.' Murhard and Kloss (who cites Panzer) give this book to Joh. Widman, whose name appears in an address to the reader; but Hain, though naming Widman, puts the work down as anonymous.

Two collections of the works of Boethius, with the Arithmetic among them, *Venice,* fourteen-ninety-one and ninety-seven, *folio.*

Deventer (Daventrie), fourteen-ninety-nine, *quarto.* 'Enchiridiom Algorismi sive tractatus de numeris integris.'

Without date, but with a letter from Balth. Licht dated fifteenhundred, *Leipzig,* printed by Melchiar Lotter, *quarto.* 'Algorithmus linealis cum pulchris conditionibus Regule detri: septem fractionum: regulis socialibus,' &c. And another, seemingly of the same date, by the same printer, 'Algoritmus linealis.'

No doubt the first English print on Arithmetic is cap. x. of 'The Mirrour of the World or Thymage of the same,' headed 'And after of Arsmetrike and whereof it proceedeth,' printed by Caxton in fourteen-eighty. (Peacock, p. 419, and Ames.)

Basle, fifteen-three.   **Nicolas de Orbelli.**   Cursus librorum philosophie naturalis venerabilis magistri Nicolai de Orbelli ordinis minorum secundum viam doctoris subtilis Scoti.' *Quarto in fours.*

This course of natural philosophy requires two pages of explanation of arithmetical, and, in particular of Boethian, terms, and nothing more. It purports to have been intended for those who were studying theology. The *Margarita Philosophica* (page 5,) seems to give the highest arithmetical limit in liberal education, and the work before us the lowest.

See page 12. John de Lortse, &c. I was perfectly justified in writing Lortse, for the name is so spelt in my copy in the clearest black letter. Nevertheless, it is John de L'Ortie, and the book is a French translation of the work of Juan de l'Ortega, mentioned by Dr. Peacock, pp. 426 and 436. Seeing de L'Ortie [sic] mentioned by Meynier, and the word thus written being nothing but the translation of L'Ortega (nettle), I compared the French work with Dr. Peacock's description of the Spanish, and found that the two must necessarily be the same. On looking very narrowly at the ſ which caused the mistake, I found in it a small hook at the bottom, which the letter ſ never has elsewhere in the work, and the letter i always has. The ſ then in Lortse is an i in which the dot (or rather small line, for so it is in the black letter) has been accidentally battered into the top of an ſ. But the alteration is so perfect that every one reads it for an ſ. This is the second paragraph which one battered letter has caused. In the new edition of Brunet it is noted that Heber's catalogue spells the word Lortse. My copy is the one which was in Heber's library.

This French translation gives Juan de l'Ortega an earlier date than the Spanish bibliographers, or those who have devoted themselves to the Dominicans, are aware of. Antonio, in his Spanish lists, and Quetif, in his Dominican one, both give Ortega no earlier date than 1534. But it appears that he was translated into French in 1515. Panzer (xi. 465, according to Brunet) mentions a work of L'Ortega, *Messina*, fifteen-twenty-two, beginning ' Sequitur la quarta opera di arithmetica et geometria.'

Florence (?), no date.   **Francisco di Lionardo Ghaligaio.** ' Summa De Arithmetica.'   *Quarto in threes* (N).

Florence, fifteen-sixty-two.   **Francesco di Lionardo Ghaligai.**   ' Pratica D'Arithmetica . . . . Nuovamente Rivista.'   *Quarto in fours.*

Two editions of the same book. The first must have been published before fifteen-twenty-three, the year in which Julius de Medicis, to whom it is dedicated as Cardinal, would have been addressed as Pope. It is a complete and advanced treatise of arithmetic, in thirteen books. The tenth book begins algebra, or *Arcibra*, the introduction of which from the Arabic seems to be attributed to Guglielmo de Lunis; Leonard of Pisa and Giovanni del Sodo are also mentioned as writers.

Venice, fifteen-twenty-six. **Joh. Fr. Dal Sole.** 'Libretto di Abaco novamente Stampato : Composto per lo excellente maestro Joanne Francisco dal sole Ingegnero : utilissimo a cadauno per imparare per se stesso senza maestro. Dimandato breve introductione.' *Octavo in twos* (O).

A very elementary work, from the school of Borgi and Pacioli, proceeding up to the square root.

―――― fifteen-forty-five. **Gaspar de' Texeda.** ' Suma De Arithmetica pratica y de' todas Mercaderias Con la 5 orden de' contadores. Hecho por Gaspar de' Texeda, Con Privilegio Imperial.' *Quarto in fours* (ON).

The colophon, which probably contains the place, is torn out. In numeration the places are distinguished throughout the first part of the book. Thus, 950777200000 is 950‖777q⁰200‖000, where ‖ (the two lines should be joined at the bottom) stands for *mill*, and q⁰ for *cuento*, a million. The proof by sevens and nines appears, as in Borgi. The work is rather more advanced than that of the latter, and proceeds to the extraction of roots. The greater part is commercial. Rules for fractions are given, not complete. The rule of three opens with an invocation of the Trinity : Sfortunati goes further, for he connects the rule with the doctrine.

London, fifteen-forty-six. ' An introduction for to lerne to recken with the pen, or with the counters accordyng to the trewe cast of Algorisme, in hole numbers or in broken newly corrected. And certayne notable and goodly rules of false positions thereunto added, not before sene in oure Englyshe tonge by the which all maner of difficile questions may easely be dissolved and assoyled Anno. 1546.' *Octavo.* At the end is : ' Imprynted at London in Aldersgate Strete by Jhon Herford.'

London, fifteen-seventy-four (printed by John Awdeley). ' An introduc- [sic] of Algorisme, to learn to recken wyth the Pen or wyth the Counters, in whole numbers or in broken. Newly overseen and corrected . . . ' *Octavo* (small) (O).

These are, no doubt, reprints of the work of nearly the same title, mentioned by Dr. Peacock (p. 419) as having been printed at St. Albans in 1537. It was a good predecessor of Recorde's *Grounde of Artes,* but it makes only a poor follower.

See page 19. In speaking of Rudolph, I had overlooked that Dr. Peacock (p. 424, note) describes his *Algebra* from inspection, as having been edited by Stifel himself in 1571.

Rome, fifteen-eighty-three. **Christ. Clavius.** 'Epitome Arithmeticæ Practicæ.' *Octavo* (small) (O). See page 33.

Witeberg, sixteen-four. **Gemma Frisius.** 'Arithmeticæ Practicæ Methodus Facilis.' *Octavo* (small). See page 25.

See page 26. Macpherson (Annals of Commerce, eighteen-five, quarto, vol. i. p. 146) gives (from Anderson, i. 409) a French edition of Stevinus on book-keeping, *Leyden,* sixteen-two, *folio,* as quoted by Anderson from a copy in his own possession. The title he gives is ' Livre de compte de prince à la manière d'Italie en domaine et finance ordinaire : contenant ce en quoi s'exerce le très illustre et très excellent Prince et Seigneur Maurice Prince d'Orange, &c.'

Frankfort, sixteen-thirteen. **Joh. Faulhaber.** ' Ansa inauditæ et mirabilis novæ artis . . . . ' *Quarto.* Same place, author, and form, sixteen-fourteen. ' Numerus figuratus . . . . ' and sixteen-fifteen, ' Mysterium Arithmeticum . . . . '

These singular medleys of arithmetic, algebra, prophecy, and nonsense, seem to be all one publication. The date of the first is JVDICIVM, written at the bottom of the title-page. Alter the order of the letters into M.DC.VVJII., and it becomes 1613.

London, sixteen-fourteen. **James Dowson.** ' De Numerorum Figuratorum Resolutione.' *Octavo.*

This work gives Pascal's table.

No place nor date. **Joh. Harpur.** ' The Jewell of Arithmetick : or, the explanation of a new invented arithmeticall Table . . . . ' *Quarto.*

Perhaps this work is the *Merchant's Jewel* mentioned by Bridges (p. 44) as of 1628, a date which, from the character of the work, I should be disposed to give it. I can find no mention nor use of decimal fractions in it. The operations are performed on a kind of abacus.

Johnson, presently mentioned, describes a ' most excellent instrument invented by Mr. William Pratt, called, The Jewell of Arithmatick ;' which is perhaps the one described in the work above, though with a different inventor's name.

London, sixteen-thirty-three. **Joh. Johnson** (Survaighor). ' Johnson's Arithmatick In 2 Bookes. The first, of vulgare Arithma : with divers Briefe and Easye rules : to worke all the first 4. partes of Arithmatick in whole numbers and fractions by the Author newly Invented. The Second, of Decimall Arithmatick wherby all fractionall operations are wrought, in whole Numbers, in Marchants accomptes without reduction; with Interest, and Annduityes.' Second edition. *Duodecimo* (O).

In his decimal fractions, Johnson has the rudest form of notation; for he generally writes the places of decimals over the figures, thus—

1.2.3.4.5.

146·03817 would be 146|03817

Otherwise, his system is tolerably complete.

London, sixteen-forty-nine. **R. B.** (Mr of Arts). 'Arithmetick Symbolical In one Book. In which the Mystery of Numeration by Symbols is revealed.' *Octavo.*

A short and easy treatise on Algebra.

London, sixteen-fifty-five. **R. B.** 'An Idea of Arithmetick at first Designed for the use of the Free-Schoole at Thurlow in Suffolk. By R. B. Schoolmaster there.' *Octavo* (O).

Decimals and algebra, all on the plan of Oughtred.

Antwerp, sixteen-sixty-two. **Jean Raeymaker.** 'Traicte d'Arithmetique Contenant les quatres especes, avec la regle de trois, et la Practique.' *Octavo.*

Not a first edition. Nothing but questions and their answers, as in John Speidell's work of sixteen-twenty-eight.

London, sixteen-seventy-five. **C. H.** 'The Golden Rule made Plain and Easie.' *Octavo.* Second edition.

No date nor place (I think about sixteen-eighty). **J. W.** (of Brandon). 'A progression in Arithmetical Progression . . . . . with Some things very remarkable in that Mystical Number 666.' *Octavo in twos.*

The author shews how to form squares and cubes by differences. He is very charitable, for his day, to the Papists: for, setting down twelve as the divine number, and that of the apostles, he seems to imply that the Roman church, as having a number made up of sixes, may be half Christian and half apostolic.

London, sixteen-eighty-four. —— 'Enneades Arithmeticæ; the Numbering Nines. Or, Pythagoras His Table extended to All Whole Numbers under 10000.' *Octavo in twos.*

The plan is, to have 99 rods instead of nine, in Napier's set; so as to have the first nine multiples of all numbers short of 100.

London, sixteen-eighty-five.   **John Wallis.**   ' A treatise of Algebra, both Historical and Practical.'   *Folio in twos.*

London, sixteen-ninety-three.   **John Wallis.**   ' De Algebra Tractatus ; Historicus et Practicus.   Anno 1685 Anglice editus ; Nunc Auctus Latine.'   *Folio in twos.*

The Latin, or augmented edition, is the second volume of Wallis's collected works.   See the Introduction to this Catalogue.

London, sixteen-ninety-seven ——.   ' Computatio Universalis, seu Logica Rerum.   Being an Essay attempting in a Geometrical Method, to Demonstrate an Universal Standard, whereby one may judge of the true Value of every thing in the World, relatively to the Person.'   *Octavo.*

The author, cutting down life to its half, to allow for childhood, sickness, &c., proceeds to a calculation how the true value of every thing is to be estimated numerically.   This, and Craig's more famous book (see Useful Knowledge Library, *Probability*,*) are consequences of the Principia.   It is curious that as soon as *force* was widely made known as an object of numerical measurement, some people began to fancy prudence, pleasure, and pain could be submitted to the same process.

Breslau, seventeen-seventy-nine.   **Joh. Ephr. Scheibel.** ' Einleitung zur mathematischen Bücherkentnis. Eilftes Stück.' *Octavo* (small).

See the Introduction.   This part contains the arithmetical bibliography; but the paging does not begin after that of the ' Zehntes Stück.'   In fact, the bibliography of this book of bibliography is a study of itself.

Göttingen, seventeen-ninety-six, ninety-seven, ninety-nine, eighteen-hundred.   **Abr. Gotthelf Kastner.**   ' Geschichte der Mathematik . . . .'   Four volumes *octavo.*

See the Introduction, as to Kästner.

---

* I have seen this book sold with my name upon the cover as the author.   To this I have no objection, except my knowledge of the fact that it is the joint production of Sir John Lubbock and Mr. Drinkwater-Bethune.

LIST OF 1580 NAMES

OF REPORTED AUTHORS, EDITORS, &c. OF WORKS ON ARITHMETIC,

INCLUDING THE INDEX TO THIS CATALOGUE.

The numbers refer to pages of the present work ; and the asterisk prefixed shews that the author has been referred to by Dr. Peacock in his History of Arithmetic.

J. H. M. A., 68.
Abercromby, 51.
Abraham, Aben Ezra, 95.
Abraham, Chaja.
Abraham of Prague.
Abt.
Adami.
Adams.
Addington.
*Adelbold.
Aeneæ.
Agrippa, 13, 21.
Ahlwardt.
Albert.
Alberti.
Aldhelm.
Alexander de Villa Dei.
Alfraganus.
*Algus, 92.
Alkindus.
Allingham.
*Alsted, 41, 42.
Ammonius, Joach.
Anatolius, 88.
Anderson, 76, 93.
Andreas.
*Andrews, Laur.
Anianus, xvi. 101.
Anjema.
Ankelin.
Apian.
*Apollonius, 76.
Apuleius, 4.
*Archimedes, 76, 84.

Archytas.
Argyrus.
*Aristarchus.
Arnold.
Artabasda (Nic. Smyrn.), 34.
*Aryabhatta.
Asclepius.
Ashby.
Astius, 88.
Atkinson, 65.
Atwood, 77.
Aub.
Augustine.
Ault, 98.
Author.
D'Autrêpe.
Avemann.
Aventinus.
Awdeley, 103.
Aylmer.
Ayres, 64.

R. B., 105.
*Bachet de Mezeriac, 47.
Bachmeister.
*Bacon, xix. 14.
Bagley, 97.
Baily, 63.
*Baker, 23, 25, 50.
Barlaam, 32.
Barlow.
Barnes.

Barozzi.
Barreme, 75.
Barres, des.
Bartel.
Bartjen, 48.
Bartiens.
Barth.
Bartl.
*Bartschius.
Bartoli.
Barton.
Bartrum, 98.
Barwasser.
Basedow.
Baselli, 77.
Bassi.
Baum.
Bausser.
Bayer.
Bayley.
Beasley.
Beausardus.
Becker.
Beckman.
*Beda, 34.
Bedwell, 35.
Behen.
Behm.
Behmen.
Behrens.
Bellie.
Benedictus, J. B., 31.
Benese, 18.
Berckenkamp.
Berger.
Berghaus.
Bergsträste.
Berkeley, 63.
Bernard.
Bernoulli, 78.
Berthevin.
Bertram.
Bettesworth, 71.
Beutel.
Beuther.
*Beverege, 66.
Bevern.
Beyenberg.
*Beyern.
Beyerus.

Bezout, 87.
*Bhascara Acharya, 86.
Bidder, 94.
Biagio da Parma.
Biermann.
Bierögel.
Bildner.
Biler.
Billy, De.
Billingsley.
Binet.
Birkin, 80.
Birkner.
Birks, 75.
Bischof.
Blagrave, 25.
Blasing.
Blasius.
Blassière.
Blavier, 75.
Bluhme.
Blundevile, 9, 30.
Böbert.
Bockel.
Bocmann.
Bode.
Boden.
Boeclerus.
*Boethius, xx. 3, 4, 5, 10, 11, 13,
    17, 100, 101.
Böhm.
Boissière.
*Bombelli.
*Bonacci (Leonard of Pisa).
Boninus.
Bonneau.
Bonnycastle, 76, 91.
*Borgi, Piero, 2, 98.
Borgo, Pietro, 2.
*Borgo, Di (Paciolus), 2.
Boscherus.
Boteler, 44.
Bouvelin.
Bourdon, 91.
Bouvelin.
Bovillus, 3.
Boyer.
Boysen.
Bradwardine, 11.
*Bragadini, 11.

*Brahmegupta, 86.
Brancker.
Brand, 63.
Brandauen.
Brander.
Brandt.
Brasser, 36.
Bredon.
Brent.
Bretschneider
Brett.
*Bridges, 44, 60, 65.
*Briggs, xxv. 42.
Brodhagen.
Brodreich.
Bronchorst.
Broschius, 43.
Brown, 62, 65, 69, 84.
Bruell.
Bruncke.
Brunetti.
Bruno (Jordano), 28.
Brunot.
Brunus.
Bubenkius.
Büchner.
*Buckley, vii. xvi. 20.
Budæus, 20.
Bullialdus, 41.
*Bungus, 28.
Bunsen.
Bunzel.
Burckhardt.
Burtius.
Buscher, 28.
Bussche.
Busse.
Bütemeister.
*Buteo, 22.
Butler, 38, 87.
Butte.
Butterman, 97.
Buttner.
Buxerius.

Cadell.
Cæsar (Jul. Patav.)

Caius.
Calandri, 1.
Caligarius (Pelacanus).
Calmet.
Camerarius, 46.
Campanella.
*Canacci.
Cap de Ville.
*Capella (Martianus).
Cappaus.
Capricornus.
Caraldus (?), 33.
Cardan, 17.
Cardinael.
Carey, 88.
Carlile, 79.
Carolus.
Carr.
Carver, 87.
Cassini, D.
*Cassiodorus.
Castiglione, 73.
Cataldus, xxi. 33, 35.
*Cataneo.
Cathalan.
*Caxton, 101.
Ceulen.
Chalosse.
Chamberlain.
Chambers, 32, 98.
Champier.
Champion, 67.
Chapelle.
Chapman.
Chappell, 80.
Charlton, 91.
Chauvet.
Chauvin.
Chelius.
Chernac.
Chladenius
Chlichtoveus, 3.
*Crishna.
Christiani.
Christofle.
Chuno.
Cirvello.
Claircombe.
Clapproth.
Clare.

Clark, 49, 85.
Clarke, 79.
Classen.
Clausberg.
Clavel.
*Clavius, 7, 33, 76, 104.
Clay, 36.
Clebauer.
Cleomedes.
Clerk.
Clermont.
Cloot.
*Coburgk (Sim. Jac.).
Cock.
*Cocker, xxii. 56, 65, 92.
Cockin, 74.
Coetsius, H.
Coggeshall, 52, 65.
Coignet.
Colburn, 94.
Cole.
*Colebrooke, 86.
Coles.
Collins, 48, 50, 51, 57.
Colonius.
Colson.
Condillac, 80.
*Condorcet, 53, 82, 86.
Copeland.
*Cornaro.
Cortes.
Cossali.
Cotes.
Counteneel.
Coutereels.
Cracher.
Crelle.
Creutzberger.
Crivellius.
Crohn.
Crosby, 80.
Crousaz.
Crowquill, 96.
Cruger.
Crumpton, 97.
Crusius.
Cuetius.
Cunn, 64, 69, 74.
Cunningham, 98.
Cuno.

Cuppaus.
Curtius.
Cusa, 10.
Cusanus, Joh. 10.

Daetrius.
Dafforne, 50.
*Dagomari.
Van Damme.
Dangicourt.
Dansie.
Danxt.
Danziger.
Daries.
Darnell, 90.
Dary, 48.
*Dasypodius.
Dauden.
Daumann.
Davenant.
Daviden.
Davidson, 71, 98.
Davies, 19, 30, 89, 95.
Davis, 86.
Dean, 97
Dechales, xv. 53.
Decker.
Dedier.
*Dee, 22.
Degen.
Degrange, 97.
Deidier.
Deighan, 83.
*Delambre, xvii. 84.
Delile, 97.
De Morgan, 89, 96.
Detri.
Deubelius.
Develay.
Develey, 97.
Dibuadius, 32.
Dicellus.
Diezer.
Digges, 24, 28, 40.
Dilworth, 75, 76.

*Diophantus, 46.
Ditton, 66, 70.
Dluskin.
Dodson, 71, 73.
Döhren.
Donn, 72.
Döppler.
Dowling, 74, 98.
Dowson, 104.
Drape.
Dubost, 83.
Dühren.
Duke.
Duncombe, 78.
Dunkel.
Dunus.
Dupp.
Düsing.
Dyer, 97.

Eadon, 97.
Eberenz.
Eggerer.
Ehrhardi.
Eichstadius.
Elend.
Elf.
Elias Misrachi.
Emerson, 78.
Enderlin.
Engelbert.
Engelbrecht.
Engelken.
English.
Eratosthenes.
Erhorn.
Erxleben.
Eschenberg.
Eslingen.
*Euclid, 63, 77.
Euler, 78.
Euteneuer.
*Eutocius, 84.
Everard, 68.
Ewing.

Faber (Stapulensis), 3, 10.
Falconius.
Famuel.
Faulhaber, 104.
Farrar, 87.
Favre.
Fayst.
Feist.
*Feliciano da Lazesio.
Felicianus.
Felkel.
Felton, 98.
Fenning, 71.
Fermat, 47.
Fernel, 8, 15.
Ferrei.
*Fibonacci (Leonard of Pisa).
Finckius.
*Fineus (Orontius), viii. 5, 15, 16, 18, 19.
Fischer.
Fisher, 34.
Fissfeld.
Fletcher.
Flicker.
Flor.
Fludd, 35.
Flügel.
Follinus.
Fontaine.
Forbes, 48.
Forcadel.
Forir, 96.
Fortia, 79.
Fortius, 16.
Forster, 38.
Foster.
Foys de Vallois.
Fraiter, 97.
François.
Fraser, 80.
Freber.
*Freigius.
Frend, 84.
Frenicle, 53.
Frey.
Friedeborn.
Friesenborch.
Friess.
*Frisius (Gemma), 25, 104.

Frömmichen.
Frommius.
Fuhrmann.
Fuller.
Funk.
Funcke.
Furit.
Fürderer.
Furtenbach.
Fustel.

Gadesby, 97.
Gaesse.
Galigai (Caligarius).
Gallimard.
Galstrupius.
Galtruchius, 50.
Galucci, 5.
*Ganesa.
Garrett, 91.
*Garth.
Gartner.
Gaudlitz.
Gauss.
Gautier.
*Geber.
Gebhardus.
Gehrl.
Geigers.
Gelder, De, 41.
*Gemma Frisius, 25, 104.
Gemma Reinerus.
Gempelius.
Gendre, Le.
Gerardus.
Gerardus (Paul).
*Gerasenus (Nicomachus).
*Gerbert.
Gerbier, 51.
Gerhardt.
Gerhard.
Gernstein.
Gersten.
Geyger.

Ghaligaio, ⎱ 102.
Ghaligai, ⎰
Gibson, 44.
Gietermaker.
Giordano.
*Girard, Alb. 26, 37.
Girtanner, 88.
Girthanner.
Glareanus (Loritus).
Gleich.
Gleitsmann.
Glogovia, Joh. de.
Glynzonius.
Gmunden, 11.
Goffen.
Goldammer.
Goldberg.
Gollner.
Good, 65.
Goodacre, 83.
Goodwyn.
Goosen.
Gordon, 97.
Gore.
*Gosselin, 21.
Gottignies, 51.
*Gottlieb.
Göttsche.
Göttschling.
Gough, 79.
Graf.
Grafe.
Graff, De.
Gräffe.
Graffinriedt.
Grammateus.
Grandi.
Graumann.
Grave.
Gravesande, S', 73.
Gray, 89, 95.
Gregory, 85.
Greig.
Gretser.
Greve.
Grey, 90.
Griesser.
Grillet.
Grossi.
*Grostete.

Gruneberg.
Grunewald.
Grüning.
Grünmos.
Grupe.
Gruson.
Grysen.
Gryson.
Guden.
Gueinzius.
Gujieno (Siliceus), 15.
Gunter, xxv.
Gunther.
Gütle.
Guy.

C. H. 105.
J. H. 68.
Haas.
Habelin.
Hagen.
Hager.
Haidinger.
Haintzelmann.
Halbert.
Halfpenny, 70.
*Halifax (Sacrobosco).
*Halley, 37, 64, 67.
Halliday.
Halliman, 52, 69.
Halliwell, 14, 15, 30.
Hämel.
Hamelius (Paschasius).
Hamilton.
Hänel.
Hardcastle.
Hardy, 76.
Harpur, 104.
Harriot, 42.
Harris.
Harrison, 94.
Harroy.
Hartmann.
Harttrodt.
Hartwell, 22.

Hartwich.
Hase.
Hasius.
Hatton, 22, 66, 67.
Hauff.
Haughton, 38.
Haupt.
Hauser.
*Hawkins, 52, 56, 58, 59, 65.
Hawney, 65.
Hay, 97.
Hayes.
Heath, 71.
Heber.
Heckenberg.
Hederich.
Hedley.
Heere.
Heilbronner, xv. 69.
Hein.
Heinatz.
Heinlin, 54.
Heinsius.
Heinzelmann.
Held.
Helden.
Hell.
Hellwig.
Helmreich.
Hemeling.
*Henischius, 33.
Hennert.
Henning.
Henrici.
Hentsch.
Hentschel.
Hentz.
Hentzen.
Herberger.
Herbestus.
Herford, 103
Herigone, 40.
Hermann.
Hermantes.
Herodian.
Herr.
Herrmann.
*Hervas.
Herwart.
Hesse.

Heyen.
Heyer.
Heynatz.
Hill, 70.
Hills (Hylles?).
Hindenburg.
Hinrichsen.
Hirsch.
Hoar, 97.
Hobelius.
Hochstetter.
Hodder, 46, 57, 58, 59, 60.
Hodges.
Hodgkin, 67.
Hoff.
Hofflin.
Hoffmann.
Hohenberg (Herwart).
Hohnstein.
Holiday.
Holland.
Hollenberg.
Holmes.
Holywood (Sacrobosco).
Holzmann (Xylander).
Holzwart.
Hood, 24.
Hoorbeck.
Hope, 78.
Horem, 11.
Horner, xxvii. 88, 89.
*Hostus.
Hotson, 98.
Howell.
Howes.
Hoyer.
Hübsch.
Hudalrich (Regius).
Huet.
Hugerus.
Hullie du Pont, L'.
Hume, 10.
Hunger.
Hunichius.
Hunt, 39.
Hurry.
Huswirt, 3.
Huth.
Hutton, 85, 89, 98.
Hylles, 31.

Hyperius.

Ihne.
Illing.
Imison.
Imhof.
Imhoof.
Impen.
*Ingpen, 36.
Ingram, 85, 88.
Intelman.
Ion.
Irson.
Isidorus.

Jackson.
*Jacob (Coburgk).
Jacobi.
Jacobs.
Jagar.
Jager.
Jahn.
*Jamblichus, 17, 46, 88.
Jaquinot.
Jarvis, 98.
Jaster.
*Jeake, 55.
Jesper.
*Joch.
*Johnson, 59, 104.
Johnston, 86.
Joncourt.
*Jones, 63, 69.
Jons.
Jordaine.
Jordan.
Jordanus, 10, 100.
*Josephus Hispanus.
Josseaume.
Joyce, 90.

Jung.
Juvigny, 97.

Kaltenbrunner.
Kampke.
Kappell.
Karrer.
Karsten.
*Kästner, xv. 106.
Kate.
Katen.
Kaudler.
Kauffunger.
Kaukol.
Kearsley.
Keckermann.
Keegan.
Kegel.
Keiser.
Keith, 78, 95.
Kellerman.
Kempten, 23.
*Kepler, S.
*Kersey, 48, 58, 73.
Killingworth.
*King, 83.
Kirby.
*Kircher.
Kirchner.
Kirkby, 67, 71.
Klinger.
Klippstein.
Klügel.
Knoes.
Knorre.
Knott.
Köbel, 9, 10, 11.
Koch.
Kochanskus.
Köhler.
Kolbe.
Königsbrun.
Königstein.
Kopffer.
Koster.

Krafften.
Kraft.
Kramer.
Krause.
Kreitscheck.
Kretschmar.
Kritter.
Kroymann.
Krüger.
Krünitz.
Kruse.
Kundler.
Küpfer.
Küster.

Du Lac.
Lacher.
Lacroix, 87.
Lagny, 55.
*Lagrange, 82.
Lambech.
Lambert.
Lamboy.
Lamotte.
Lamy, 53.
Landase.
Landus.
Lange.
Langens.
Langius.
Langner.
Lantz.
*Laplace, 82.
Lardner.
Latomus.
Launay.
Laurbech.
*Lauremberg, 36, 41.
Lauterborn.
Lauxius, 10, 100.
Lavus.
Lawson, 49.
Lax, 11.
Leadbetter, 68, 70.
Lebez.

Leblond.
Lecchi.
Lechner.
Leeuwen, Van.
*Leibnitz.
Leichner.
Leiste.
Leistikov.
Lembke.
Lemoine.
Lempe.
Lenz.
*Leonard of Pisa (Bonacci Fibonacci), xviii. 1, 17, 94, 95.
Leotaud, 45.
Leri.
*Leslie, xvi. 89.
*Leupoldus.
Levera.
Levi.
Leybourn, 46, 49, 50, 52, 54, 72, 89.
Libri, xvii. 93, 95.
Licht, 101.
Lichtenberg.
Lidonne.
Liebe.
Lieberwirth.
Liedbeck.
Liesganig.
Lindenbergius.
Lindner.
Lintzner.
Lipstorp.
Liset, 50.
Little, 80.
Littmann.
Liverloz.
Loche.
Locher.
Lohenstein.
Loher.
La Londe.
Longomontanus.
Lonicerus.
Lorenz.
Loritus (Glareanus).
L'Ortie, } 12, 102.
Lortse, }
Lory.

Lösch.
Loscher.
Lossius.
Lostau, 67.
Lotter, 101.
Lovell.
Löw.
Lowe, 70.
Lucar, 28.
Lucas.
Lucas à St. Edmondo.
Lucius.
Luders.
Ludolff.
Ludwig.
Lunenzius.
Lunesclos.
*Lunis, Gul. de, 102.
Lunze.
Luthardt.
Luya.
Lydal.
*Lyte, 36.

J. R. M., 62.
Machin, 74.
Maes.
Magelsen.
Mainwaring.
*Mair, 56, 77.
Malapertius.
*Malcolm, 66, 69.
Maler.
Malham.
Malleolus.
Mandey, 54.
Manning.
*Manoranjana.
Mantaureus.
Maphæus.
Marcellus.
Marchetti.
Mariani.
Markham.
Marleen, 97.

Marsh, 69.
Marshall.
Martin, 68, 69, 73.
Martini.
Martino.
Martinus (Herbipolensis), 101.
Maslard.
Massard.
Masterson, 29.
Mather.
Matheson, 97.
Mauduit.
Maurolycus, 24.
May.
Mayer.
Maynard, 80, 83, 95.
Mayne, 48.
Mazeas.
Mazzonius.
Meanus.
Medler.
Meenenaer.
Megillus.
Meierus.
Meinck.
Meissner.
*Mellis, 22, 27.
Melrose, 88.
Mengolus.
Menher, 23, 33.
Mercator, 98.
Merckel.
Mereau.
Meres.
Merliers.
Mesa.
Messen.
Metius, 41.
*Metrodorus.
Metternich.
Meurerus.
Meursius.
Meyer.
Meynier, 34.
Mezburg.
*Mezeriac (Bachet).
Michaelis.
Michahelles.
Michelsen.
Micraelius.

Micyllus.
Middendorpius.
Milns, 80.
Minderlein.
Mirus.
Mittendorf.
Mognon.
*Mohammed ben Musa, 93.
Mole.
Molineux, 79.
Möller.
Monhemius.
Monson.
Monzon.
*Moore, 43, 49, 52, 59.
Morin.
Morisotus.
Morland, 47.
Morrison, 97.
Morsshemius.
Morssianus.
Moscopulus (Emanuel).
Mose.
Moser.
Moses.
Mosley, 98.
Mülich.
Müller.
Mullerus.
Müllner.
Mullinghausen.
Munnoz.
Muris, J. de, 3, 4, 11.

Nabod.   }
Naiboda. }
Nagel.
Napier (Neper).
Nefe.
Nelkenbrucher.
Nemorarius (Jordanus).
*Neper, xxiii. 35.
Nesius.
Neudorffer.
Neufville.

118 LIST OF NAMES OF AUTHORS.

Neumann.
Newton, 46, 73.
*Nicholas Smyrnæus Artabasda, 34.
Nicholson.
Nicolson, 97.
*Nicomachus (Gerasenus), 17, 46, 88.
Niese.
Nisbet.
Nomer.
Nonancourt.
Nonius (Nunez), 23.
Nottnagel.
Norfolk, 15.
Norry.
*Norton, 27.
Noviomagus, 18, 34.
Nunez, 23.

Obereit.
Ochler.
Odo.
Oelse.
Oeser.
Oesfeld.
Oesterle.
Ofenlach.
Ohm.
Olaus.
*Oldcastle, 28.
Opuntius.
Orbelli, 102.
*Orontius (Fineus), viii. 5, 15, 16, 18, 19.
*Ortega, 102.
L'Ortie, J. de, 102.
Oser.
Ostermann.
Ostertag.
Osterwald.
Otto.
*Oughtred, xiv. xxii. xxv. 37, 38, 49.
Ouvrard.
Ouvrier de L'Isle.

Overheyden.
Ozanam.

W. P., 30.
Pabst.
Pachymeres, Geo.
*Paciolus (di Borgo), xx. 2, 95.
*Pajottus.
Palmquist.
*Paolo dell' Abacco.
*Pappus, 85.
Pardon, 68, 69.
Parent.
Parker, 93.
Parisius.
Parrot.
Parsons, 63.
Partridge, 42, 51.
Pascal.
Paschasius (Hamelius)
Pasi.
Paterson.
Patrick.
Pauer.
Paulinus.
Paulinus à St. Josepho.
Paulus.
Paze.
Peacock, xvii. xxiv. 14, 91, 97.
*Peele, 28.
Pelacanus.
*Peletarius, } 22, 25.
Pelletier, }
Pelicanus.
Pell.
Penkethman, 41.
Penrose, 97.
Da Perego.
Perez.
*Pergola.
Perry, 98.
Pesare, Leon. de.
Pescheck.
Petersen.
Petreius.

Petri (Nicol.), 31.
Petvin.
Petzoldt.
Peucerus.
Peurbach, 11.
Peverone, 21.
Peyrard, 84, 86.
Pfaff.
Pfeffer.
Pfeiffer.
Pflugbeil.
Philipp.
Philippus Opuntius.
Philips.
Phillips, 97.
*Philo.
Philoponus.
*Photius.
Pickering.
Pierantonio.
Pieterson.
Pike.
Pisani.
Piscator.
Planer.
*Planudes.
Platin.
Platz.
Playfair, 81.
Playford, 72.
Plotinus.
Poeppingius.
Poetius.
Polenus.
Poppin.
*Porphyry.
Porter.
Pöschmann
Postellus.
Potter.
Pratt, 104.
Prestet, 54.
Preston, 88.
Prexendorffer.
Printz.
Pritchard, 90.
*Proclus.
Prosdocimo di Beldoman o.
Psellus, 21.
*Ptolemy, 84.

Purdon, 87.
Purser.
Putsey, 98.
Pütter.
*Pythagoras, 1, 4.

Quensen.

Rabus.
Rademann.
Raeymaker, 105.
Ram.
Ramsbottom, 97.
*Ramus, 29.
Randall, 74.
Ranzovius.
Raphson, 66, 74.
Rawlin.
Rawlyns.
Raymaker.
Rea, 65.
Recorde, xxi. 21, 22, 59, 66.
Rees.
*Regiomontanus, xiv. 16, 91.
Regius, Hudalrich, 16.
Regneau.
Rehmann.
Reich.
Reiche.
Reichel.
Reimann.
Reimer.
Reinau.
Reiner.
Reinhard.
Reinhold.
Reinholdt.
Reisch, 4.
Reiser.
Remer.

Remmelinus.
Renaldini.
Renaldus.
Renterghem.
Resenius.
Ressler.
Reuss.
Reyher.
Reymer.
Reynaud, 87.
Reyneau.
Reynolds, 97.
Rhaetus, 15.
Rheticus, xxi.
Rho.
Riccius.
Richards, 97.
Richardson.
Richter.
Riederus.
Riege.
Riese, 19.
Rigaud, 19, 76.
Riley.
Ringelbergius, 16.
Ritchie, 20, 88.
Rivail, 97.
Rivard.
Rivet, 74.
Robertson.
La Roche (Villefranche), 12.
Roder.
Romer.
Romanus. }
Roomen, Van. }
Rommel.
Ronner.
Roscher.
Roselen.
Rosen, 93.
Rosenkranz.
Rosenthal.
Rosenzweig.
Roslin.
Rossel.
Roth.
Rouquette.
Le Roux.
Rouyer.
Röver.

Royer.
Rozensweig.
Ruchetta.
Ruddiman, 68.
*Rudolff, 23, 104.
Ruffus, 13.
Rühl.
Ruhmbaum.
Ruinellus.
Russell.
Rust.
Rutherford, 95.
Ryff, 9.

R. S. S., 62.
*Sacrobosco (Holywood, Halifax), xix. 13.
Sadler, 75.
Salfeld.
*Salignac.
Salmasius.
Salomon, 75.
*Sanchez.
Sarret.
Sauer.
Saul.
*Saunderson, 68, 71.
Savary, 98.
Savonne.
Saxon.
Scalichius.
Schäffer.
Scharf.
Schatzberg.
Schedel.
Scheffelt.
Scheibel, xv. 106.
Schellenberg.
Schelliesnig.
Schessler.
Scheubelius, 20.
Schey.
Schiener.
Schiller.
Schirmer.

Schlee.
Schlegel.
Schleupner.
Schlönbach.
Schlosser.
Schlügel.
Schlüssel.
Schmalzried.
Schmid.
Schmidt.
Schmotther.
Schneidt.
Schonberger.
Schoner, 16, 29, 35.
Schoop.
Schöttel.
*Schott, 45.
Schramm.
Schreckenberger.
Schreckenfuchsius.
Schreyber.
Schreyer.
Schröter.
Schübler.
Schuler.
Schulze.
Schultze.
Schumacher.
Schürmann.
Schurz.
Schwarzer.
Schweighäuser.
Scoten, Van.
Scott, 97.
Seck.
Seckgerwitz.
Segner.
Segura.
Selden.
Seller.
Sempilius.
Seriander.
Servin.
*Severinus, Boethius.
*Sfortunati, 16, 103.
Shakerley.
Sharpe.
Shelley, 73.
Shepherd.
Shield.

Shirewood, 10, 101.
Shirtcliffe, 69.
Siegel.
Silberschlag.
Siliceus, 15.
Simpson, 71.
Sinclair.
Sinner.
Siverius.
Smart, 63.
Smeeth, 97.
Smith, 72.
*Smyrnæus (Nich. Artabasda),
    34.
Smyters.
Snell, 64.
*Snellius, 32.
Soave.
Soc. Usef. Kn. 93.
Soc. Chr. Kn. 98.
Sole, Dal, 103.
Sosen, Von.
Sotter.
Speidell, 37.
Speusippus.
Spies.
Spiess.
Spitzer.
Splittegarb.
*Srid'hara.
Stadius.
Staps.
Stapulensis (Faber).
Starcken.
Stark.
Starke.
Steger.
Steinius, 25.
Steinmetz.
Stenius.
Stephanus à St. Gregorio.
Stephens.
Sterk, 16.
Stern.
*Stevinus, xxiii. 26, 32, 35, 104.
Steyn.
Sthenius.
*Stifelius, 19, 23.
Stigelius.
Stillinger.

M

Stimmingen.
Stock.
Stöffler, 8.
Stonehouse, 72.
Storr.
*Strachey, 86.
Strauchius.
Stricker.
Strigelius.
Stritter.
Strozzi.
Strubius.
Strunze.
Sturmius, 62.
Suevius (Sigism.).
Suisset, 12.
Sulzbach.
Sutton, 76.
Svanberg.

Tabingius.
Tabouriech.
Taccius.
*Tacquet, 56, 63.
Tacuinus, 34.
Taf.
Tait.
Tallen.
Tangermann.
Tannstetter, 11.
Tap, 33.
Tarragon.
*Tartaglia, 21.
Tassius.
Tate, 85, 96.
*Taylor, 76, 86, 87, 90, 97, 98.
Tedenat.
Telauges.
Telfair.
Tennulius, 46.
*Teruelo.
Tessaneck.
Tetens.
Texeda, 103.
*Theon, 41, 84.

Theophrastus.
Thevenau.
Thierfelders.
Thoman.
Thompson, 80, 98.
Thornycroft.
Thorpe, 97.
Thoss.
*Thymaridas.
Timaus.
Tinwell, 91.
Tissandeau.
Tocklerus.
Toissonière.
Tollen.
*Tonstall, xv. 13.
Torchillus (Morssianus).
Trapp.
Treiber.
Trenchant.
Treu.
Treyrerens.
Trincano.
Trommsdorf.
Trotter, 80, 85.
Turner.
Tylkowski.
Tyson, 91.
Tzechani.

Udalrich (Regius).
Ulman.
Ursus.
Ursinus.
*Urstisius, 24.

J. B. V., 84.
Valla, 3.
Vallerius.
Vandamme.

Vandenbussche, 23.
Vandendyck.
Vandervelde.
*Vasara.
Vayra.
Vega, 89.
Veiar.
Velhagen.
Vellnagel.
Veronensis.
Vicar.
Vicecomes (Visconti), 24.
Vicum.
Vierthaler.
Vieta, xxi. 42.
Villefranche.
Vincent.
Vinci (Leonardo da), 20.
Vinetus.
*Virgilius Salzburgensis.
Visconti, 24.
Vitalis.
*Vlacq.
Voch.
Vogel.
Voigt.
Volck.
Vollimhaus.
Vomelius.
Voster.
Vries.
Vulpius.
Vyse, 81.

J. W., 43, 105.
Wagner.
Walbaum.
Walgrave.
Walker, 86, 90.
Walkingame, 80, 96.
*Wallis, xiv. xxii. 37, 38, 44, 69,
    82, 85, 106.
Walrond.
Walther.
Waninghen.

Warburton, 81.
*Ward, 65, 69.
Waserus.
Wastell, 63.
Watson, 98.
Weberus.
Webster, 40, 68.
Wechselbuch.
Wecke.
Weddle, 89.
Wedemeier.
Wehn.
*Weidler, 66.
Weigel.
Weinhold.
Weise.
Wells, 55, 64.
Welpius.
Welsh, 73.
Wencelaus.
Wendler.
Wentz.
Werz.
Wesellow, 36.
Westerkamp.
Weston, 68, 73.
Wetterhold.
Whiston, 73, 74.
White, 93, 97.
Whiting.
Wicks, 97.
Widebergius.
Widman, 101.
Widmann.
Wiedeberg.
Wiedemann.
Wiesiger.
Wigan.
Wilborn.
Wilcke.
Wilder, 74.
Wildvogel.
Wilhelmus.
Willemson.
Willesford.
Williams.
Williamson, 77.
Willich.
Willichius.
Willsford.

Wilson, 69.
Windssheim.
Wing, 53.
*Wingate, 42, 44, 48, 59, 73.
Winterfeld.
Wirth.
Witekind.
Witt, xxiv. 33.
Wittmer.
Wochenblatt.
Wohlgemuth.
Woit.
Woldenberg.
Wolff, 68, 70.
Wolphius.
Woltmann.
Woodward, 97.
Wördenmann.
Worley.
Worsop.
*Wright, 69, 97.
Wucherer.
Wurffbain.
Wurster.
Wurstisius.
Wybard, 43.

Xenocrates.
Xylander, 21.

Youle, 86.
*Young, 89, 90, 93.

Zaake.
Zabern.
Zacharias.
Zahn.
Zaragosa.
Zell.
Zeller.
Zeplichal.
Zesen, Van.
Zinge.
Zoëga.
Zon.
Zonsen.
Zubrod.
Zucchetta.
Zurner.

THE END.

LONDON:
PRINTED BY LEVEY, ROBSON, AND FRANKLYN,
Great New Street, Fetter Lane.

Printed in the United States
By Bookmasters